Adventures of the
Yorkshire Shepherdess

Adventures of the
Yorkshire Shepherdess

AMANDA OWEN

MACMILLAN

First published 2019 by Macmillan
an imprint of Pan Macmillan
20 New Wharf Road, London N1 9RR
Associated companies throughout the world
www.panmacmillan.com

ISBN 978-1-5098-5267-3

1 3 5 7 9 8 6 4 2

A CIP catalogue record for this book is available from the British Library.

Typeset by Palimpsest Book Production Limited, Falkirk, Stirlingshire
Printed and bound by CPI Group (UK) Ltd, Croydon, CR0 4YY

Visit **www.panmacmillan.com** to read more about all our books
and to buy them. You will also find features, author interviews and
news of any author events, and you can sign up for e-newsletters
so that you're always first to hear about our new releases.

To the one I love

Contents

Introduction

When gathering the sheep from the moors I have often looked down onto Ravenseat, sitting as it does in its little hollow, and been 'tekken' with the aura of contentment that envelops the farm that I have called home for two decades. It isn't the neatest or most orderly of places, but it exudes a warmth that is both heartening and welcoming.

The first time I came here, what struck me was the sense of quiet. It's so peaceful, but the place is rich in history, having seen so much life during its near thousand-year existence. The labours of people from times past were plain to see when I looked across the partly cobbled yard towards the timeworn buildings all surrounded by a chaotic frame-work of crumbling drystone walls. In its heyday, a hundred and fifty years ago when manpower and horsepower ruled, nearly a hundred people lived at the top of Swaledale – now it's thirty, and that includes my brood. In the eighteenth century, Ravenseat was a small hamlet with eleven families in residence. For weary travellers passing through, refreshment for both body and soul were available at the public

1

house (which is now our farmhouse) and at the Inghamite chapel (which is now our woodshed). For the residents of far-flung settlements like Ravenseat work was either to be found in the coal and lead mines or on the many small farms, but the decline of the mining industry in the late nineteenth century led to a mass exodus with two thirds of the population of Swaledale leaving to find employment elsewhere. Farming suffered too, the smaller farms becoming less viable. Some were completely abandoned, and the land amalgamated to form bigger enterprises. Slowly but surely the lifeblood of the dale trickled away, leaving behind only isolated farmsteads and derelict mine workings, the relics of bygone times.

I first visited Ravenseat on a dark October night in 1996. I was a contract shepherdess in my early twenties and had been asked to collect a tup (ram) from a farmer called Clive Owen. Clive was single, in his early forties, and like me had a passion for farming and the great outdoors but was not so bothered about home decor. It's fair to say the farmhouse was a wreck, with damp carpets, black mould and wallpaper peeling from the walls. It was less than inviting and, with nothing in the way of heating other than a small coal fire in the living room and a temperamental range cooker in the kitchen, the house felt dank. Clive, though, was funny and easy-going and we became friends and then something more.

I moved in and gradually – very gradually, I should say – over time, the furnishings and fittings have been upgraded but there's still plenty of room for improvement. I was mindful that a working farmhouse must be, in the first instance, practical. I couldn't guess the number of times I've had lambs warming beside the hearth, or being bathed in the kitchen sink, or had to step over a recumbent calf on the fireside rug. For me, this is the essence of a farmhouse, not a highly polished Aga or a Cath Kidston apron. Ravenseat is sparse where it needs to be, with bare flagged floors that can be scrubbed, but also decidedly cluttered places where items often needed in a hurry are stacked and ready to hand. For example, on the overmantel in the kitchen there are a couple of bottles of calcium and mixed minerals ready for use when we are presented with an emergency case of ovine grass staggers, a commonplace metabolic disorder that occurs when there are low levels of magnesium in a sheep's blood. The medicines are more effective and work quicker if warmed to blood temperature, so although they're not the perfect visual adornment for the overmantel, it is the ideal storage place. A couple of pot dogs might look more decorative but are not as useful.

After Clive and I had been a couple for four years, I finally proposed.

'Does ta think we should get married?' I asked him.

'Mebbe.'

'Does that mean yes?'

'I suppose so.'

Granted, it wasn't the most romantic of proposals.

We married at St Mary's Church in Muker in July 2000. Today Ravenseat is home to Clive, myself and our nine children, plus terriers Chalky, Pippen and young pup Sprout, a whole host of sheepdogs, an amorous peacock, too many hens to count, three horses and an aged pony, a small herd of cows and about 1,000 sheep. Throughout the summer months we have guests staying in our shepherd's hut and if that doesn't keep us busy enough, we have a stream of customers wanting to enjoy the al fresco afternoon teas that we provide during the walking season. Most of our visitors are travelling from the west to the east on the arduous 192-mile Coast to Coast walk which will take them two weeks to complete. They will have walked for a week by the time they reach us, and although Ravenseat lies on one of the shorter stretches of the journey – it's just twelve miles from Kirkby Stephen to Keld – the terrain is challenging. The climb out from Hartley takes them up to the Nine Standards Rigg, standing at 2,172 feet above sea level, where nine sizeable ancient stone cairns command the summit. The view east is over the bleak moors of Swaledale, and to the west you can see the green plains of the Eden Valley. From here on it is a precarious path down Whitsundale, through knee-high wiry ling (heather), scrub and peat haggs, with walkers needing to sidestep the bog moss or find their feet sinking into treacherous ground.

'T'as me beat why thoo would wanna live at such a godforsaken spot,' was the comment I got from a farmer out for a drive one abysmal wet November afternoon, but I love this place and the challenges it brings. I never want to live anywhere other than Ravenseat but the simple fact is that we are tenant farmers. Ravenseat is part of an estate and has been for centuries; it doesn't belong to us. That insecurity and the question of what the future would hold for our family kept me awake at night, wondering, worrying.

'Life's too short to worry about the what ifs, yer not

supposed to know what's in't future,' Clive would say.

'Aye,' I'd agree, 'but unfortunately the fact that I'm considerably younger than you means I'm gonna be the one left with the worries.'

He'd scowl – nobody likes to think about their mortality and Clive was no exception. Then he'd turn the conversation to his advantage: 'An' this mi dear is why yer need to look after mi, cater for mi every need, an' whilst we're on that subject I'd like a cup o' tea, I'm parched.'

Gradually the idea of buying our own place in our beloved Swaledale took root – a permanent home for our family, somewhere that we could call our own. The story of our search starts in 2013, at which point we had six of our nine children. Raven, our eldest child, was born in 2001. As the oldest she takes on much responsibility for her younger siblings and is a great help to me. She has a wise head on young shoulders; she's practical and down to earth but also academic, and is now in sixth form, studying Chemistry, Biology and Maths for A level. She says she wants to be a doctor, but sometimes it's a vet, other times a research scientist, so who knows? I tell all the children that they can be whatever they want to be, it's down to them.

In 2003 Reuben came along. He is our resident engineer and is most likely to be found in a shed tinkering with something mechanical. He's a tall, handsome lad but always streaked with oil. Deft with a socket set and accomplished with a welder, he's at his happiest when confronted with a seized engine or malfunctioning machine. The price of metal has fallen so low that the scrap man doesn't come up to Ravenseat very often, so everything I put onto the scrap heap somehow migrates back into the farmyard to be used for another of his restoration projects. He's now got four vintage tractors, which he is always fiddling with, and is first to show up with a spanner if anything needs fixing around the house or farm. He's dyslexic and, much to his chagrin, has to have 'assisted learning' at school but he's excellent at concentrating if it's something he's interested in. Give him a book about tractors and engines and he'll read that happily enough. He put a whole new engine into our skid-steer loader and it worked all winter, although I do know that the success of the project all hinged on a piece of wood, carefully crafted to the right size, holding a rubber hose in place. He's also extremely good at talking to people and has many friends, both young and old, the common denominator being a passion for machinery. It is not unusual for Reuben to introduce us to one of his friends who are perhaps themselves 'vintage' in terms of age but who are willing to impart vital knowledge and knowhow relating to tractors and stationary engines.

Miles, who was born in 2006, is like Reuben in that he doesn't like school, but unlike his older brother he's discovered a new interest, girls; he's a real ladies' man. With his wide brown eyes peeping out from under his foppish fringe, he is bound to charm. He loves animals, nature programmes and the countryside. Every year we replenish our ageing flock of hens with end-of-lay hens destined for processing. Miles takes them under his wing, caring for them in the barn until they become acclimatized to the weather at Ravenseat and can become free-range.

He also has his own flock of sheep, a dozen Texdales

(Texel/Swaledale crosses). He lambs his little flock himself; everything is timed very carefully to make sure that his sheep's due dates coincide with the Easter school holidays. He gets to see it all, and he was heartbroken when one day he went outside to feed his sheep and found one had escaped and got into the feed store, overindulged, become bloated and died. It's entirely his flock and he makes the decisions about who he keeps and who he sells with just a hint of guidance from Clive. He is quiet, sensitive and knowledgeable and lives for the day when he can farm full-time.

Edith came along two years after Miles and she too is enthusiastic about farming. Like Raven she is studious, more so than the boys, and is doing very well at school. But she's happiest outside in her boiler suit and wellies and loves to come with me to the moor to shepherd the sheep. When Clive was preparing his beloved tups for Tan Hill Show in the spring, it was Edith who spent hours tonsing them – plucking out stray white hairs – when the rest of the children would give up, losing patience and interest, after about half an hour.

She's also fascinated by the birds that we have up here. She watches over the nests during the breeding season, and knows exactly who's who. Her familiarity with the resident birds is quite astounding, but not really surprising given the amount of time she spends out in the fields. I suppose it is simply about knowing your patch, knowing the lie of the land, and watching the mannerisms of the birds. You soon get to identify and understand what is right there in front of you. My RSPB observer was astounded by her photograph album which consisted of lots of selfies with lapwing, moorhen, oyster catcher and curlew nests. She even kept a note of how many chicks hatched.

Violet was born in 2010, and she is a real tomboy, always nursing some war wound. A scrape down one leg, a huge blood blister. She and Clive are locked in competition about their injuries: whatever happens to Violet, Clive has always had it worse (or better, from their point of view). Clive's blood blister covered the whole of the palm of his hand, so he won that one. Violet is always pushing the boundaries of what is possible, climbing higher up a tree than anyone else dare or jumping into the river from a loftier ledge. One summer's day she dragged the plastic slide across to the river bank, climbed up it and aquaplaned across the water like a skimming stone whilst everyone else looked on. She's completely fearless and thinks nothing of being Reuben's human guinea pig for whatever contraption he has crafted – latterly this was a home-made snowboard that was impressive in its speed but not so much in its manoeuvrability.

She has an arty side and spends many a happy hour making clay pots and figures. She discovered her own little claypit on the moor amidst the peat haggs and heather, and can often be found there digging up muddy clay, squeezing it and kneading it, nails blackened and hands grey. She busies herself making models of ladybirds and hedgehogs, as well as bowls and cups. Some of her handiwork she brings home to 'fire' in the black range oven and then decorate, but she leaves a lot of them there, in situ, to dry out in the sunshine. I have on occasion been quizzed by a coast-to-coast walker who has chanced upon this little home-made shrine of pot animals amongst the heather.

Sidney arrived in 2011. With a cheeky smile, dancing eyes and freckles he is a live wire with an enquiring mind and is an excellent source of real-time commentary on daily events at Ravenseat. He is wise beyond his years owing to spending so much of his time in Reuben's company. Both are mechanically minded but Sidney's enthusiasm, coupled with his idolization of his eldest brother, has led to him being somewhat taken advantage of, as he tends to be used as a gofer. 'Sid, gofer this; Sid, gofer that.' And he does so, willingly, proud to be helping out. Wherever Reuben is, you'll find Sidney is not so far away.

I hope you enjoy reading about what happens next at Ravenseat, as our youngest three children arrive, along with a stoical plumber, a travelling monk, an auctioneer and a very lucky calf. Chalky behaves badly, Little Joe the Shetland pony defends his honour, and we face some of our toughest battles yet in the worst summer and winter we've ever encountered.

1

Goin' to the Chapel

I was running late. This in itself was not unusual. However well-organized I'd tried to be there always seemed to be another little thing that needed to be done before I drove out of the farmyard. I am a creature of habit, and with six children to care for as well as a farm to run I had to cover every eventuality. But that day there was no excuse for it, the hospital appointment had been at the forefront of my mind for weeks. Northallerton was my destination, where I was due at the maternity outpatients' unit for a twenty-week scan.

'There's pasties in the bottom oven and milk's in't fridge,' I said to my long-suffering husband Clive. 'You'll have to let the horses into their stables at darkening if I'm not back.'

'Listen up,' he replied. 'I's quite capable of looking after mi self an' everything 'ere will be fine.'

I knew he was right, it was just that leaving the farm was a wrench. I regarded Ravenseat as my piece of heaven, a place far from the madding crowd, and the long winter months always accentuated this feeling of splendid isolation. Perhaps

I was subconsciously trying to delay my inevitable journey into civilization.

'Have you got yer phone, Mand? I'll be talkin' to yer.'

I nodded. Our one mobile phone was used by every member of the family from Reuben upwards. It didn't work at Ravenseat, or even anywhere in the vicinity, so we took it with us only if we were venturing out of the dale.

The silly thing was that I hated rushing, and on such a glorious morning it seemed completely wrong that I'd have to focus entirely on the road ahead rather than be able to take in a few of the sights and sounds of Swaledale in springtime. It was March 2013 and the land was just beginning to waken after the most brutal of winters, the worst we'd experienced in a long time. It had tested us both mentally and physically but now the snow had finally receded, the white blanket reduced to ever-shrinking pockets that lingered in the cold shadows of distant crags. Lifeless greys and umber tones had dominated the land-scape for what seemed like forever, but now the moors were looking more verdant. The clumps of seaves (rushes), their stalks now softening and greening, and a carpet of mosscrop with white cotton-puff heads bobbing in the gentle breeze, were the first signs of the better things yet to come. Lambing time was just around the corner, the hardest work in the shepherding calendar, but also a time of new life and hope.

'An' for gawd's sake pick mi up summat to read,' Clive had shouted as I finally flung a few recyclable shopping bags into the back of the Land Rover. 'If I 'ave to read that flamin' tup catalogue yance more I think I'll ga' mad.'

'*Practical Parenting* magazine perhaps,' I replied, smirking. 'There's usually a few of them in't waiting room.'

I watched him manhandle three-year-old Violet and toddler Sidney towards the farmyard. I don't think that he heard – or if he did, then I got no reaction.

I drove slowly out of the farmyard, dodging a discarded child's bicycle and avoiding a couple of chickens that were enthusiastically scratching away at a clart of mud that lay on the concrete track. I drove steadily through the ford, the crossing place a few feet upstream from the simple but ancient packhorse bridge that has spanned Whitsundale Beck for many centuries. The heat from the early-morning sunshine had warmed the water and steam rose from the slowly flowing beck. My wheels made gentle waves, and a startled moorhen hurriedly skated the surface and disappeared beneath the overhanging greenery of the riverbank where the water lapped. I glanced at the clock; I had an hour and a half to get to the hospital, fifty miles away, and reckoned that I could just about make it. I switched the radio on, twiddled with the tuning dial, then promptly switched it off when all I got was the crackle of white noise.

I glanced across into our Close Hills pastures, sure that I could see a sheep, not so unusual of course but it certainly wasn't supposed to be there. All the yows were still on their heafs – open areas of moor where they quietly grazed – held there not by physical boundaries, walls or fences but by their own natural homing instinct. They would only be gathered down into the fields when the lambs became due. I looked away. Today at least I was determined not to get too embroiled in what was a constant battle of wits between myself and my little woolly wanderers. Today it was all about me and the new baby.

I slowed a little as I passed Bridge End, where my friend Rachel lives. Meg, her old sheepdog, was snoozing on the doorstep. Sometimes, after I'd rounded the corner before the Keld Youth Hostel, I'd stop, wind down the window and take a few photographs of the valley known as Crackpot, but not today. Not only was I rushing, but there was something different in my line of vision. There in the field opposite the Wesleyan Chapel was a 'for sale' sign. Another had been nailed to a wooden batten across one of the building's rotting window frames. I mentally noted the name of the estate agency and kept going – I would look up the details of the property later as for now I had other things to think about. The children, who were wise to the fact that there was soon to be a new addition to the family, had been arguing as to whether they'd be getting a brother or a sister. As it stood, there were three of each: three girls, Raven, Edith and Violet; and three boys, Reuben, Miles and Sidney. Even Clive had been wondering.

'A tup or a gimmer?' he'd asked one morning as I refused my usual morning cup of tea. I'm an avid tea drinker, so the moment I found the mere thought of a brew distasteful we knew that a baby was on the way; it was as accurate an indication as any pregnancy test.

I didn't want to know the sex of the baby before it was born, it didn't matter one bit to me, but either way I couldn't please everyone.

'I'd like a sister,' said Raven.

'I'd like a brother,' said Reuben.

'I'd like a puppy,' said Edith.

I could drive on autopilot, I knew that road so well. Drystone walls ran either side, uneven and roughly built in the higher reaches, gradually becoming more uniform and aesthetically pleasing as the road meandered its way gently down the dale. After thirty miles the road opened out and the fields were replaced by the sprawl of suburbia. I passed Richmond with its castle and bustling market square surrounded with resplendent Georgian townhouses. From here onwards the hills and dales gave way to flatter, fertile countryside and finally to red-bricked housing developments and industrial estates as we neared Northallerton.

At the hospital I lay on the bed, wellies dangling over the end, while the sonographer ran the scanner over my tummy.

'Baby's dad not with you today?' she said, not looking up, her eyes on the screen that was turned slightly away from me.

'Nah,' I said. 'He's at home looking after t'lal' uns. They'll be on wi't sheep right now.'

'Ahh,' she replied, now typing on the computer keyboard. 'You won't be helping out at all now, I suppose.'

I didn't reply. I didn't want to fib, and yet I didn't want the truth – the fact that I was still very much hands-on and fully intending to be for the duration of my pregnancy – to be mistaken for carefree complacency. I decided to change the subject.

'Everything all right?' I asked.

'Absolutely fine,' she replied, tilting the screen towards me so that I could try and interpret the grainy image on the monitor.

'I can't make that out at all,' I said. I was just as hopeless at deciphering what was on the screen when our sheep had their ultrasound scans. She pushed the probe harder into my stomach.

'There,' she said. 'Now you can see the baby's—'

'Nooooo, don't tell me,' I said, averting my eyes. 'I don't want to know if it's a he or a she.'

Subconsciously, though, I had decided now that it was a baby boy.

She printed off a little picture, nothing too telling, just something to show Clive and the children.

I made my way back to the car park and rang home. Nobody answered, but that wasn't a surprise. Clive would be outside. I sat for a while in the Land Rover, looking at the little picture, then decided that I may as well take a walk down the High Street and check the estate agent's window to see if the chapel's details were up yet. They weren't, so I went in.

'You're quick off the mark,' said the bespectacled lady perched cross-legged on a stool in front of a computer. 'The sign only went up this morning.'

She rifled through the reams of paper strewn across the desk, finally finding the information for me. I thanked her and walked back up the street. I was nearly back to the car park when my phone rang.

'You all right?' Clive said.

'Yep and I've got summat to tell yer.'

'NOT TWINS!' he said.

I reassured him that there was only one baby, that all was well, and that I'd show him the scan picture when I got back home.

'It's about the chapel at Keld,' I said excitedly.

'Please tell me that you haven't found God,' he muttered.

I explained that I'd seen the 'for sale' sign.

'How much?' he asked.

'Erm,' I said, cocking my head to one side, the phone wedged between my shoulder and my ear. I removed the paper from my pocket and unfurled it. 'Eighty – eighty thousand it says.'

'Hmmm, we'll talk about it later,' he said.

Later that afternoon we sat together at the kitchen table.

'Is ta sure that we want a chapel, Mand?' Clive said as he studied the particulars on the sheet.

'No, 'course I'm not sure, but I think it's worth a look,' I said. 'Properties in this neck o' t'woods come on t'market so infrequently it'd be daft not to think about it.'

'There's no 'arm in 'avin' a look I suppose,' he conceded as he supped his tea.

'I'll ring 'em, arrange a viewing,' I said. 'Right now.'

The appointment, a ten-minute slot, was scheduled for the next day. We arrived to find a line of cars parked outside; it seemed that every man and his dog was coming for a look. Never had chapel been so busy; a congregation had gathered, not to worship but with a view to buying it. Violet and Sidney came along, Clive too – in body, anyway, as he didn't seem to be there in spirit. In fact, I got the distinct feeling that he was just humouring me.

We went to the door but didn't go in. The estate agent, wielding a clipboard and a flashlight, was clearly busy showing another potential buyer around. The children were happily splashing in a small puddle whilst Clive hopped from foot to foot with impatience. I stepped back and looked hard at the stone edifice. It certainly could not be described as eye-catching. If anything, it was plain, the simplistic matter-of-fact architecture a reflection of the values that the Wesleyans held dear. I could see why Clive was underwhelmed, and when, finally, we were taken for a tour of the building's interior, it did nothing to enhance his mood. It was dark inside because the windows had been boarded up, but where part of the ceiling had collapsed, tiny chinks of light were visible – the gaps between the slates in the exposed roof. The crumbled ceiling plaster and damp lats now lay mouldering on the parquet floor. A bare bulb hung suspended by fraying fabric flex, a length of baler twine having been used to anchor it to a cross-member. The agent noted my interest in the electrics.

'No problem with utilities – mains-connected water and electric – although it will need rewiring.'

'No kidding,' Clive muttered quietly under his breath. I flashed him a dirty look. The inside was in a state of severe disrepair, a far cry from my last visit there in 2002 when I had been invited by Jimmy Alderson to take baby Raven to be blessed at a Sunday service. I remembered sitting there with Jimmy and Elenor and a squirming Raven dressed in an embroidered white christening gown and lace bonnet. The altar remained, along with a lectern and line after line of pitch pine pews, all covered with a fine layer of dust. Some of the pews were damp, the wood beginning to rot where rainwater had dripped onto them. The agent, perhaps sensing that Clive wasn't buying into the idea of making the place into a home, began his spiel.

'There's no planning permission on it, is there?' I asked, interrupting him.

'I wouldn't see there being any issues in it being granted,' he replied. 'The chapel's not listed.'

I nodded. My idea – and everybody else's too, I assumed – was to convert the chapel into a house. It would be no simple feat: it was a sizeable building, and it didn't particularly lend itself to being made into a cosy home.

'Huge potential,' said the estate agent. 'Structurally, it's quite sound.'

'With a bit of imagination and a decent architect, we could do something with it,' I said, looking around as the agent shone the flashlight here and there, never holding the light for too long on anywhere too unpleasant.

'We'd need bottomless pockets, more like,' said Clive. 'The walls are knackered.'

'It'd make a fantastic atrium, with a vaulted ceiling and a spiral staircase perhaps,' said the estate agent.

'Floors are knackered too, or as good as, and the windows are shot,' Clive retorted. We were on a downward spiral!

'The arched window would make a fantastic statement piece,' said the agent, casting his arms wide. 'You've got space, lots of space. What about a minstrels' gallery?'

My mind was racing along now – I'd definitely be borrowing a copy of *Homes and Interiors* magazine the next time I went to the doctor's. I'd already decided that I wanted to preserve the integrity of the building, keep some of its heritage alive. If anything, I wanted to make more of a statement; the Wesleyan simplicity was far too boring for me. I wanted more in the way of religious symbolism. I was soon awakened from my daydream.

'Damp,' said Clive. 'The spot's bloody damp.'

'Will yer stop being so negative,' I said. The agent looked uncomfortable, like he didn't want to be caught in the middle of a domestic. Clive, sensing my frustration – and in an effort to prove that he wasn't totally uninterested – set off to have a look at the back of the outside of the building to check out the drainage and the general lie of the land. I reckoned that the real reason for his departure was that he'd spied farmer Rukin's tup hoggs grazing in the field behind. It was purely coincidental, he said.

The children were getting bored, Sidney had become a bit clingy, and what with my burgeoning tummy and him perched on my hip, I was beginning to tire.

'C'mon, it's time we were all gettin' back,' I said. 'We've got stuff to do.'

We'd fed and bedded up all of the animals in the farmyard but there was still the moor sheep to feed. It was at this point that Clive came back and surprised me, for, quite out of the blue, he said to the agent, 'How much is ta wantin' for it?'

The agent, who had started to lose hope, quickly perked up.

'Erm, eighty thousand pounds is the list price,' he replied.

'Well, we'll 'ave it then,' Clive said.

My jaw dropped; I hadn't expected that and now I felt scared. We didn't have that amount of money, I needed to find a way to raise it, and I'd never had a chance to enquire about a loan or a mortgage. Fortunately, the divine intervention that would buy me some time swiftly arrived.

'Erm, I'm afraid I can't actually sell you the chapel right now. There is considerable interest in the property and there's a chance that someone might offer the vendor, the church that is, more money.'

'Well, I don't get that,' said Clive. 'If that's 'ow yer gonna play it then yer should be auctioning it off.'

And with that, we left. It only took ten minutes to get back home and frankly that was a good thing as we ranted for much of the journey. We both were of the opinion that if something had a price then that was how much it was.

'Don't get yerself upset, Mand, ga' an' talk to the bank and then we'll put in an offer an' see what 'appens.'

'I really didn't think you wanted to buy the chapel, Clive,' I said. 'Yer couldn't find owt good to say about it.'

'First rule of buying anything: don't look too keen or interested, play it cool,' he said, wisely. 'The ball is firmly in their court, they know that you want it, you're just going to have to wait and see what happens.'

Patience is not something that I am blessed with but, with lambing time being upon us and making preparations for the new baby, there was plenty to keep me occupied. Lambing time is all encompassing; for weeks on end we'd live, breathe and talk about almost nothing other than sheep. But news travels fast and I was sure that if the 'for sale' sign was replaced with a 'sold' sign then I'd get to know and, as the saying goes, no news was good news.

One lunchtime we were sitting discussing the merits and drawbacks of feeding the animals on fodder beets, and whether chopping the root vegetables with a spade would make the cows either more or less likely to choke on them, when the postman appeared at the door with a bundle of envelopes and papers.

'Hurray, some reading material,' said Clive. 'I ain't seen a paper in weeks.'

Most of the envelopes were familiar – feed bill, vet bill and suchlike – only one was not identifiable. I opened it first.

'What's that?' asked Clive, glasses perched on the end of his nose as he peered over the top of his *Farmers Guardian*. It was from the estate agency. Inside was a preprinted form and a covering letter. The letter stated that the chapel was still on the market and that we should, if interested, submit our best and final offer and proof that we had funds available to pay for it. It also stated that because the owner was the church, a charitable organization, they would take into consideration people's reasons for wanting to buy.

We filled in the form there and then – eighty-two and a half thousand pounds was as much money as we could give, and if that was not enough then it was too bad. I included a bank statement and letter from the bank manager agreeing to the loan of the remaining money required. I went to town on our reasons for wanting to buy it: being local, having an ever-increasing family, being tenant farmers, and the fact that I wanted to preserve much of the chapel's heritage during the renovation. In a last-minute thought before I sealed the envelope, I included that my eldest daughter Raven had been blessed there. The deadline for putting in the offers was just a fortnight away, so once again I had to be patient and wait and see what transpired.

There were rumours abounding as to who had bought the chapel long before we got the official letter telling us we'd been unsuccessful in our bid. Maybe it really was destined to be a nightclub, petrol station or fast-food drive thru; frankly I didn't care. I tore up the letter in a fit of pique and threw it onto the fire. Then I quietly seethed as I waited to find out who really had beaten me to it. It didn't take long until word got out – they'd sold it to a couple from London who were going to convert it into two holiday rentals, and they'd got it for just five hundred pounds more than I'd offered.

To say I was annoyed was an understatement. I spent all of ten minutes composing an angry email to the estate agent outlining the effort we had put into securing loans and gathering up paperwork but how in the end it had all come down to the money, when supposedly locality and links to the chapel were meant to be influencing factors. I told them that I was formally withdrawing my original offer, and should they ever find themselves in need of a buyer then I would give them fifty-one thousand for it and not a penny more. After signing off I added a PS that the vendor, the church, should also remember the tenth commandment, forbidding greed and the amassing of worldly goods, then I clicked the send button and felt immediately better.

I don't suppose that anyone read it, not properly anyway. Perhaps just a cursory glance told them that it was another disgruntled customer; I'm sure that I wasn't the first or last. Months later we heard that the deal had fallen through; planning permission had not been granted as there was not enough off-road parking for two holiday cottages. I felt quite smug, like maybe they regretted not selling the chapel to us, but now with our new baby girl Annas – a surprise in that I was sure I was expecting a boy – I was focused on other things and had lost interest in the property dream. For over a year we'd hear of new buyers and the subsequent failed deals. I did wonder whether they'd ever concede defeat and accept my paltry offer but nothing materialized and, frankly, I was glad, for by now other unexpected opportunities had arisen.

Firstly came the sale of Smithy Holme, a derelict farmhouse set back off the road, perched in a commanding position above a wooded hillside overlooking the steep Keldside pastures. It was here that our old friend Tot Haykin had lived since a boy, farming the few acres that surrounded it. The house and surrounding buildings, though structurally sound, had fallen into disrepair. The house needed gutting and starting afresh. Back in the early sixties when electricity had finally come up to the very top of Swaledale, a payment of a thousand pounds had been asked of every household beyond Bridge End, of which there were not many. A few of the farmhouses like Ellers and Birkdale had been abandoned years before, and these dwellings – although now renovated and lived in – remain off-grid to this day. Smithy Holme was bypassed too, Tot and his parents finding the cost of this newfangled electricity to be just too much. And so, the pylons were erected, lines strung tantalizingly close to Tot's house, and the lights in Upper Swaledale were switched on. Gaslamps, candles and hurricane lanterns were consigned to the past, except at Smithy Holme where time stood still. Tot was a lifelong bachelor and lived with his mother for many years until her death. When, finally, age got the better of him, he left the house and moved into sheltered housing at Gunnerside further down the dale. He'd return every day to tend his few sheep, his furthest field being adjacent to ours at the bottom of the Close Hills pastures. I'd often see him leaning against the wall beside the tumbledown sheepfolds just watching his flock. From a distance he cut a lonesome figure, his gaunt frame always swathed in countless layers of loose clothing and a deep-pocketed sheepskin coat reminiscent of a market trader. Beneath the dirty threadbare tweed flat cap that cast a shadow over his angular face, his eyes danced with mischievousness. I enjoyed talking to him, he'd recall times past, of his parents and friends who were now long gone, wild parties and bad winters when 'all was happed up wi' snaw'.

Inevitably the day came when Tot became too old and ill to come to his sheep. Friends took over his daily jobs and for a little while Tot was kept updated on the goings on whilst in a hospital bed. Sadly, weeks later he died. His sheep were sold and the land too; all that remained was to find a new owner for Smithy Holme.

It was a place that we knew very well, we'd often walked there on the footpath from Ravenseat which was, until the middle of the last century, the main route both to and from our house. As footpaths often do, it followed the gentle contours of the hillside, meandering through the pastures and then skirting the edge of Cotterby Scar before joining the road to the nearest village, Keld, an hour's walk away. The children loved to poke around in Smithy Holme's outbuildings as Tot had, over the years, accumulated what can only be politely described as a bit of junk. This included a canoe, an old generator and the skeletal remains of what Reuben reliably informed me was a Land Rover series I.

These carefree walks became more purposeful when the 'for sale' sign went up. We now went walking with a view to buying. I liked the haphazardness of the outbuildings, the stunted tree with tangled roots that grew through the foundations of the low garden wall which divided the frontage and what I would call the home field. The house itself was square and angular, solid and well built, and where before we'd looked wistfully upon it as being the home of an old friend, we now had to open our eyes to the realities of the place as a potential renovation project.

There was no denying that to take on a house in this state would be a big undertaking and, after a great deal of soul-searching and a few reality checks, we were confronted with the harsh truth that we didn't have the skills necessary to be able to do the work and definitely not the time required to get the project off the ground. Heavy-heartedly we accepted that once again this was an opportunity that must pass us by.

I couldn't believe that in such a relatively short space of time two properties in the upper dale had come onto the market and each time it hadn't worked out for us.

'Bide yer time,' Clive would say. 'Summat'll 'appen.' And once again he was absolutely right.

It was September 2013, a beautiful summer holiday had been enjoyed by the children and now our focus was on making ready for Muker Show. Every year we would prepare our show sheep, bake cakes and buns, and enjoy a day away from the farm meeting up with friends whilst the children overdosed on sweets and ice cream and the bouncy castle. It was often the case that you'd catch up with friends that you hadn't seen for a whole year since the last show, and this was exactly how we finally came to find our perfect house.

Susan and Graham had lived at The Firs for the thirty years that Clive had farmed at Ravenseat and we knew them well. It would be fair to say that we were neighbours, as only a couple of miles separates our dwellings as the crow flies, but where Ravenseat had a council-maintained road leading right down to the packhorse bridge, The Firs did not. Susan, like myself, had a love of horses, particularly native ponies. When we bumped into each other whilst out and about we'd talk about our equine adventures and often mishaps. When I had first moved in with Clive, he had accepted that my horse was moving in too and considered it a small price to pay for a jodhpur-clad girlfriend. Of course, as time went on, I accumulated more horses and inevitably the jodhpurs didn't fit quite as well as they once did but Clive, although not a horse lover himself, didn't complain . . . well not too much and not enough to make any difference.

As soon as Raven was big enough, I bought her an old gelding called Boxer at a riding-school sale. He was a pony of indiscriminate breeding, thickset, chunky and already long in the tooth when he came home with me to Ravenseat. He came equipped with an antiquated English leather saddle and a bridle with a severe, double-pelham bit as poor Boxer's mouth had been deadened after years of being pulled by the hands of novice riders. His days of hard, monotonous work were now behind him, he had the chance to kick up his heels and enjoy a leisurely semi-retirement. I'd take him out on a little hack occasionally but for the most part he was tasked with the very important job of carrying Raven in a basket saddle whilst I either led him from another horse or on foot. He put on weight, his spiky rubbed mane and tail became lustrous again, and during the summer months his coat developed a sheen. His crumbling hooves no longer needed to be shod, I dressed them myself with a rasp and tripod, another purchase at the riding-school sale. I was proud to win first prize with him in the veteran class at Trainriggs Show at Kirkby Stephen. One of the other exhibitors at the show remarked that she'd learned to ride on Boxer almost twenty-five years ago and that he must now be well into his thirties.

Susan told me that when Raven required a mount of her own to ride independently, she had just the perfect pony. Little Joe was a Shetland that she had bought for her granddaughter who had now outgrown him. The day that Little Joe was scheduled to come to Ravenseat was the first and only time that I had actually been to The Firs itself and, as it happened, I arrived home empty handed, for, after jolting my way down to the house in the Land Rover, Little Joe had stubbornly refused to go into the trailer, leaving Susan with the task of walking him to our house. The road down to The Firs was little more than a narrow rutted track, steep in places, which cut through two gated fields and disappeared from sight in a final tight bend before sweeping around and stopping abruptly outside the house. I had no real lasting memory of the place and Clive had never actually been there.

Joe came to us on loan over a decade ago and has since introduced Raven and seven of her siblings to the delights of being in the saddle. There is not a bad bone in his body, he doesn't bite or kick, is never sick or sorry, but has a tendency to be stubborn and an ability to escape from anywhere at any given time.

Clive hates him. Or rather he says he hates him, calls him a demon and blames him for every gap that appears in the drystone walls.

'Bugger's bin scrattin' his arse on t'wall.'

It is true that Little Joe has supersonic hearing and can detect the sound of a feed bag rustling at a thousand paces. There is no denying that after homing in on wherever Clive was feeding his flock Joe could desecrate the whole pastoral scene by galloping around, sending sheep flying this way and that. He'd then eat their cake, totally ignoring Clive's loud protestations. Many times, Clive would return to the yard, red in the face, and announce that Little Joe's days were numbered and if he got colic from eating all the sheep food then he absolutely deserved it and there'd be no vet coming to save him.

Little Joe, of course, never ailed a thing, but now age has crept up on him, his face has greyed, his hip bones jut out and his back has dropped. He gets extra rations, a cosy stable every night and recently I have even seen Clive throw an extra armful of meadow hay over his stable door before bedtime.

Clive and I were enjoying the show and the September sunshine, listening to the silver band playing in the bandstand. Wintertime was far from our thoughts, but not Susan and Graham's, it seemed, as they had a favour to ask.

'We were wondering if you'd mind keeping an eye on our place during the winter,' they said.

''Course not,' I said. 'Are you off to sunnier climes? A holiday somewhere?'

'Well, sort of,' said Susan. 'We're going to Kirkby Stephen for the winter, so not a holiday as such, we've bought a house there.'

They had reluctantly decided that it was time to leave their beloved house and move to civilization, nearer to amenities and services and where everyday life was less of a challenge. They'd even found some land nearby for their horses.

'We'll come back and spend one last summer there and then put the house on the market.'

Year on year of hard winters, trudging through the snow up and down the track, cut off from the world and struggling to keep warm, had taken its toll. Still, it was going to be a wrench for them, leaving their rural idyll behind, I could see that.

'It'd be really good to know that nothing has frozen up, that the water's still running. We'll be back in the spring when the weather has picked up.'

When we called on them for instructions and the keys, Clive and I were shown around the house, and we noted where the stop tap, fuse box and all the storage heaters were. After writing down their new contact number, we left, promising to return regularly to check on things.

As it happened, that winter was not a hard one; it did snow, and it did freeze but not for any great length of time. We dutifully called in from time to time, switching on heaters when required whilst hopefully not accruing the absent owners a sizeable electricity bill. Not long after they were back in residence we met on the road whilst out walking with the children.

'The house is up for sale,' said Susan. 'Just on the estate agent's website though, I couldn't bear to put a sign up at the top of the track by the road.'

It made sense, in a way. They'd spent all those years living privately and quietly and I suppose the thought of nosy people invading their oasis was unbearable. Then she said it.

'I don't suppose that you'd be interested in buying it?'

Obviously, I had thought about it, not as a realistic idea but more of a pipe dream. I needed a wreck and one that was priced accordingly, a renovation project that I could work on as and when I could afford it. The Firs was too good for me.

'I'd love to, Susan,' I said, 'but it's just not the right time for me, I don't have a healthy enough bank balance to even consider it.'

'That's a real shame,' she said, 'because I would have liked you and your family to have had it.'

I told her that unfortunately I couldn't and that I was sure that she would find a buyer – it was, after all, a perfect rural retreat. There was also another reason why house-hunting was off the agenda, one that I'd kept quiet about. I was now pregnant with my eighth child.

Once I was back at home, I went online to have a look at the house particulars: 'A rare opportunity to buy a traditional farmhouse and land in Upper Swaledale.' I scrolled through the text. 'With period features dating back to the 1640s,' it trumpeted. There were photos of the interior, but far more of its big selling point: its views and location. Only a stone's throw from the farmhouse flows the River Swale. Online there maybe contradictory articles stating that the Swale begins at the confluence of Whitundale Beck (that flows through Ravenseat) and Sledale Beck just below Hoggarths Bridge, and that what flows past The Firs is merely a tributary. But I'm more inclined to believe my old friends like Jennie, who was born at Ravenseat and has lived in Swaledale all her life – she says this stretch of water is already the youthful beginnings of the mighty River Swale, that the Swale comes into being a little further upstream where Birkdale Beck and Sledale Beck meet. The situation of The Firs is magnificent, with shelter from the weather and prying eyes, and it's nestled in tightly between the rising bronzed moors and a copse of ash trees. Immediately surrounded by its own fields and not bothered by any near neighbours, it looked like paradise. Unfortunately, its price reflected this desirability. Over half a million quid! The Yorkshire Dales was a popular place for second homes and, with only a limited number of houses and a seemingly infinite number of potential buyers with a lot of money, it seemed that once again my dream would be thwarted. It was an outrageous price for a farmhouse that would once have been classed as a lowly dwelling, a place where peasant farmers would eke out a living from the poorest of land. I vowed not to look again.

Time flew by, and we, too, found ourselves putting a property on the market. Robert, Clive's eldest son from his first marriage, moved out of our ground-floor flat at Kirkby Stephen where he'd lived for over a decade. We had used the proceeds from the foot-and-mouth compensation in 2001 to buy the flat, and Robert had lived there whilst working for us at Sandwath, the other farmland and buildings that we rented for a while at Kirkby Stephen. Now that Robert had taken over Sandwath himself, and got a girlfriend, the time was ripe for him to move into a bigger place, and we were left with a small flat to sell. We thought this would give us some savings to put towards a housing project of our own, should another opportunity arise, but not being spacious nor in the best of locations the flat languished on the market. Every so often the estate agent would ring and inform us of a viewing, but nothing ever materialized.

'If you wait long enough the right buyer will come along,' said the agent. 'You have got to be patient with these things.'

It seemed that Susan was having no luck with selling her house either and she too had been issued with the same advice. There had been a few viewings of The Firs, we knew.

'A handful of dreamers,' she said when I met her while she was out walking her dog one afternoon. 'Folks who wouldn't last two minutes when the realities of life in the rural backwaters really sank in.'

'And the others?'

'The worst sort,' she said, looking defiant. 'Them that want to modernize it, iron out the lumps and bumps and the charm of the place. They tell us how they're going to make our home *habitable*! Can you imagine? Can't you buy it, Amanda? You like things as they are supposed to be.'

I smiled; I did, as she said, like things as they were supposed to be, and places that had retained their true heritage were thin on the ground. I repeated that, sadly, I was in no way able to buy it.

'Skinted, not minted!' I said and reiterated that it would be my dream to be able to own The Firs but that it was out of my league. Later, I told Clive that I'd been talking to Susan.

''Ave they getten that 'ouse sold yet?' he said. 'Or are we lookin' after it for another winter?'

I told him that they, too, were having difficulty finding a buyer, although for different reasons than ourselves.

'I don't think Susan likes what the potential buyers want to do with the house, and she's trying to put them off,' I said. 'I think Graham is humouring her, he knows how much the place means to her.'

'Mmmmm,' Clive said, not really listening.

'She still wants us to buy it,' I said.

'I wonder if they wanna part-exchange it for a flat at Kirkby Stephen?' he said jokingly.

'I doubt it,' I said, 'they've already bought a house there and what would they want with a pokey little flat anyway?'

I thought that it was a ridiculously stupid idea; there was no way that I was going to put that suggestion to Susan, ever.

'We should ga an' see 'em, tha knows,' Clive said. 'Sort this lot out once an' for all. It's no good Susan putting folks off thinking that if she waits lang enough we're going to buy it.'

He was right, it needed sorting out once and for all.

We arranged to go over at the weekend when Raven was at home to keep an eye on the littler ones. Susan and Graham welcomed us with tea and biscuits, and I couldn't help but wonder whether they thought that we had good news for them. I'd decided that Clive should do the talking but just to make sure there were no crossed wires, I thought I'd be clear from the outset.

'We 'aven't won the lottery,' I said.

'An' we 'aven't sold the flat,' said Clive, sipping his tea.

'I don't suppose you'd consider selling the flat to us, like a part-exchange?' Graham asked.

I nearly choked on my tea; I could not believe my ears, my despondency evaporated. I had a complete turnaround from despair to elation.

'I'd like to do it up as a holiday let,' he continued, 'a project, something for me to do.'

'Let's see if we can make this happen,' Clive said, looking rather pleased with himself. He stood up and shook Graham's hand.

Next came reams of paperwork, to and from solicitors and estate agents, surveyors and the bank. Nothing is ever simple, even when all parties are in agreement.

Now I had only one more person to convince that buying The Firs was a good idea: my bank manager, Lionel. It was make-or-break time – his input (lending us most of the money) was critical, so I needed to persuade him that it was a serious business proposition and that I would be able to recoup the money by letting The Firs for holidays. Lionel offered to come out to meet us at Ravenseat and see first-hand what we did and what we planned to do, to look at our business plan and projections and go through formalities. Needless to say, I was terrified and had no idea what I should do to impress him so opted for plying him with tea, freshly baked scones and cake, then chaperoned him down to the shepherd's hut so he could see my flourishing bed-and-breakfast project. There was no let-up in visitors and it helped our cause greatly that Clive was kept busy temporarily running the business of serving afternoon teas whilst I was being grilled. The doorbell kept ringing as customers put in their orders and Clive scurried back and forth with loaded tea trays.

Next, I took Lionel to The Firs and set about sharing my vision with him as he followed me from room to room. What he probably didn't realize was that I was convincing myself as much as I was trying to convince him and scaring myself silly at just how much work needed doing in not so much time. We needed to rent out The Firs as a holiday let as soon as possible to pay back what we owed, but in order to do this we had to renovate it sympathetically.

For all my enthusiasm it was still a big leap into the unknown, a project of immense magnitude – and not only that, I was hiding the fact that I was pregnant with Clemmie, our eighth baby, for fear of sounding completely feckless.

'Have you any plans for expanding your family?' Lionel asked me, as tactfully as he could.

'No, no plans to,' I said, crossing my fingers behind my back and breathing in. After a couple of torturous hours, Lionel finally gave me his verdict. 'I am going to lend you the money and I have no doubt that you will succeed with this project,' he said as he shook my hand. I welled up – to this day I don't think that he knew what his belief meant to me, for without his go-ahead the whole project was a non-starter.

No matter what lay ahead, this was a defining moment. We had committed to buying ourselves a house that would become our family home one day, a place that we could love and where we could finally put down some permanent roots.

2

A Family Home

There was no ceremonial handover of ownership, it was more of a gradual process as Susan and Graham slowly moved their belongings out of The Firs. I felt for Susan. I recalled packing up and leaving my little cottage at Crosby Ravensworth. Even though I was happy to be moving to Ravenseat to be with Clive, my husband-to-be, the whole upping sticks and closing the door for the last time still hurt. How difficult it can be to dispose of items collected over the years. I had books that I had never read, a belt that didn't fit anymore (truthfully, it was a little on the tight side when I bought it as a teenager at Pink Cadillac, the most fashionable shop in Huddersfield), spanners that had belonged to my father, and mouldering notebooks and diaries. My house clearance had been a mere fraction of what Susan and Graham had to deal with, but I knew that it was the reminiscing that took the time and that emotions would be running high. I had other things on my mind too, namely the newest member of the family, baby Clemmie. She had arrived early and – reminiscent of times past – had been born on the hearth rug in front of the open fire at Ravenseat.

Clive and I would wait until after tea and then drive down to The Firs to see how much progress had been made with the moving out. It wasn't that we were wanting to move our things in, more that we needed to see the blank canvas and decide what was the most logical starting point for the renovation. When that day finally came, we realized there wasn't a logical starting point. To see the place stripped, devoid of its clutter, furniture and warmth, made me shudder. Clive and I wandered from room to room in silence while the stark emptiness lay bare the realities of the task ahead of us. It was extremely daunting: what had initially seemed like a relatively simple spruce-up had snowballed into a monumental undertaking. And we couldn't let the house sit empty for too long. I needed to recoup some of the money that I had invested in order to pay back the bank. The house needed to earn its keep.

'Oh, what have I gotten myself into,' I said to Clive. 'How am I ever gonna get this fit for people to stay in?'

'It's not that bad,' said Clive. 'Imagine what you'd 'ave 'ad to do if we'd bought the chapel at Keld or t'Smithy Holme.'

'We wouldn't 'ave been skint an' we could 'ave gotten an architect – it'd 'ave been a blank canvas,' I said.

However, when the sun was setting over Ash Gill and the house, ivy clad and sturdy, was basked in a gentle mellow glow, I was filled with hope and joy. This was a place that, like Ravenseat, had everything: centuries of history, stories of people who had lived and worked here, and above all a peaceful sense of intimacy. There were, unfortunately, the sodden, grey days when the dwelling appeared sullen and seemed to afford little comfort when you were inside its dank walls. The cold seemed to permeate the very core of the place.

The furnishings were replaced with sparse oddments that we brought from Ravenseat: a few kitchen chairs, an electric heater for whichever room we were working in and a couple of clipping stools that we'd found in a barn. These doubled as either saw benches or tables. Clive and I set about making the downstairs sitting room into somewhere that we could retreat to at lunchtime. Susan had left a dolls' house, a game of bagatelle and a few jigsaws, and these kept Sidney and Annas amused in one corner. Little Clem was happy in her travel cot or car seat.

The one saving grace in this room was that it had a sizeable French potbellied stove, and if there was one thing that we were not short of it was fuel to keep her lit. Chewed architrave and skirting boards, rotten door frames and floorboards, the house kept us in firewood if nothing else.

Clive and I weren't half bad at demolition, excelling in the slash-and-burn technique, but as the jobs stacked up it became abundantly clear that we were going to need expert help. As farmers, we liked to think that there wasn't much that we couldn't turn our hand to but, although we did possess many basic skills, we were woefully underqualified when it came to electrics, joinery and plumbing.

On finer days, at the weekends and during the school holidays, the children would all come down to The Firs. When the weather was pleasant, they would head off to explore the surrounding barns and fields. It was only five acres but still it was a new playground for them and they ran riot.

'Should we be worried?' Clive would say when we hadn't seen hide nor hair of them for an hour or so.

'Nivver worry,' I'd say, 'they'll be back wi' some tale or other.' And they would eventually appear. Dishevelled and dirty, talking of wading through streams and climbing trees, they'd empty their pockets of the grubby mementos picked up en route. You could say that they surveyed the whole place for us, leaving no stone unturned. They knew where the gaps were in the drystone walls, where the stone land drains ran and where the demolished barn foundations were that could supply us with extra walling stone.

Reuben and Miles in particular got to know the lie of the land and became very useful groundsmen, as tidying up outside was of the utmost importance. Roll after roll of fencing wire had been used to divide up the fields into smaller paddocks but I wanted to revert to the original drystone-wall boundaries. It was important that the fields were stockproof as we did intend to take Miles's small flock of Texdales down there to graze, but there was no need to have the place resembling Colditz with coils of wire bridging gaps and suspended over gutters. It all had to go, and so the boys were armed with fencing pliers and an empty bucket in which to put the staples. Soon we had even more fire-wood, namely fence posts, all surplus to requirements. I took my chainsaw down and sawed them into logs. The children gathered up a whole tractor-load of ugly wire, rolled it up as best they could and put it, and two cast-iron baths that had been used as water troughs, out for Daz, the scrap man. Reuben set about making a little wicket gate from some of the reclaimed wood, which allowed people to walk down to the riverbank, whilst Miles cleared a nettlebed that grew on an old muck midden with a scythe.

The rich, black soil from the muck midden and rubble from previous building projects were all removed and redistributed where they could do good. I hesitate to use the word 'landscaping' to describe this work, as it was undertaken by Reuben in his vintage mini dumper. Everything was done in fits and starts; some days we could get the farm jobs done quickly and spend the majority of the day at The Firs, on other days we would run out of time altogether and vow to get to work down there the next day.

Clive is a very good drystone waller, and he has had lots of practice over the years but more in a practical sense, repairing walls, rather than constructing an object of beauty. The lawn at the front of the house morphed into the field and was crying out for a retaining wall that would separate the two areas and neaten up the frontage of the house. It was hardly a building project where he could let his imagination run wild, but still it was one that he felt perfectly at home with, and one that he could create entirely from the stone reclaimed from the field itself. The recycling or robbing of building materials is nothing new. When I'd go along gapping with Clive, we'd occasionally find stones that had been decoratively hewn long ago for other purposes. Door jambs and mullions from long-forgotten dwellings, and stones intricately chiselled by someone with an eye for detail and precision. To handle such a stone would always fire my imagination. Who were these early stonemasons, these skilled men who worked with the most basic tools but left us with a tantalizing glimpse of what went before? A gate stoop, a large angular stone with iron hanging still intact, was incorporated into Clive's new wall, as was a find of Miles's: a fossilized tree root that resembled a snake skin. There were no plans regarding the wall, it just evolved as load after load of stone was unearthed and delivered via the quad bike and trailer.

'I's gonna build a seat in this wall,' said Clive, and he did. Not the most comfortable of seats but a place to perch, nevertheless. When, finally, the last topstone had been set, we all stood as a family and looked in admiration at Clive's handiwork.

'It looks like it's allus bin there,' said Raven.

That was a compliment indeed, and exactly what we hoped to achieve with the rest of the house.

We were aware that there was only a certain amount of landscaping work that could be carried out with a spade and the mini dumper. The boys had made great strides, but it was time to bring the old orange digger down. This was music to Reuben's ears. It had been inherited from our friend Alec who had used it for renovations when he was landlord of Tan Hill, England's highest inn and our nearest pub. Now he'd retired, and just concentrated on his other passion of sheepdog trialling, he had no need for a digger. We were the grateful recipients of this slow-moving rust-bucket, and Reuben was delighted.

'Can I drive it there?' he said excitedly. 'Pleeeeeease.'

Raven rolled her eyes. 'It'll take you aaaaaaages,' she said.

'Seriously, you want to drive it all the way from t'Ravenseat to t'Firs.'

He nodded vigorously.

After he agreed to drive it cross country through the fields, Clive gave him the nod. It was only a two-mile journey, but it took him all day. First the digger overheated and boiled, then one of the tracks came off.

The rest of the children, who were playing in the farmyard at Ravenseat, provided a running commentary as the orange outline crawled along at a snail's pace. When Reuben finally reached his destination, he described all the folks he'd waved to, who'd passed by the fields in cars, apparently none of them batting an eyelid at a young boy quietly driving a digger along.

'I tell yer what though,' he said, 'I think I saw mi' geography teacher frae school an' he fair glowered at mi.'

'Thoos done well, Reubs,' Clive said, patting him on the head affectionately, 'but now it's time that I took back the controls of yon digger.' And with that he slowly reversed it down to the bottom of the lawn, swung around the boom and knocked over one of the two coniferous trees that grew in the corner.

'Look what you've done to mi bloody tree! I can't believe you've just done that,' I shouted; I was furious, Reuben thought it was funny.

'Ey, Mum, yer gonna 'ave to change name o' t'house now – it's not The Firs anymore, you've only got one fir tree now, so it's The Fir!'

We had already decided that we were going to plant more trees; we wanted as much greenery as possible. There was a wooded copse where a cuckoo could be heard calling in the springtime and a few magnificent mature ash trees grew alongside the river, their roots exposed in the crumbling eroded riverbank. I was hopeful that where we had failed with tree planting at Ravenseat, owing to waterlogged heavy soil, we might succeed at The Firs, where the soil was deeper and richer. Maybe in a few hundred years a whole line of trees would grace the skyline, but for now I was going to have to content myself with finding a replacement fir. This time it was eagle-eyed Miles who spotted, out of the school-bus window, a sad-looking Christmas tree amongst the firewood and pallets of a recently constructed bonfire.

'Are yer sure it's got its roots on?' I asked as we discussed it at the tea table.

'It's in a pot, Mam, it must 'ave.'

We didn't regard this as stealing – it was obviously unwanted – so we scaled the mountain of chipboard, pallets and branches and came home triumphant with the tree. I don't know that it's possible to hold great affection for a tree, but that one certainly came close to being loved. Edith cradled it in the back of the Land Rover for the journey home; Sidney and Violet tenderly unravelled its coiled roots from the plastic pot before carefully placing it in its ready-dug hole. Even now the children often go and check on its progress, recalling how they rescued it from its fiery fate.

To be at The Firs is to experience quiet, a silence that can seldom be found in the busy lives that we lead. Ravenseat, too, is a peaceful place, but its tranquil, sleepy appearance belies the bustling activities that are part and parcel of a working hill farm. Barking dogs, the cattle in the barn or tups in the stables all contribute to the sounds that surround us. One day Edith and I were alone at The Firs working in the garden, trying to ascertain what was a weed and what wasn't, when we heard distant voices, children's voices. We both stopped what we were doing and listened hard.

'Did you hear that?' I whispered to Edith. She nodded.

'Violet, weren't it?' she replied.

We both stood quiet for a while but heard nothing more. I was not entirely convinced that my children's voices could travel all the way from home, over a distance of two miles or so, but it is true that on a still day – as this was – when there is not a breath of wind, the cries of birds on the wing can carry over a great distance. On countless occasions I've heard the haunting call of the curlew or the drumming of the snipe and yet seen just the tiniest speck upon the horizon.

I've also had call, on occasion, to reprimand Clive for his expletive-filled rants at Bill the sheepdog. Having apologized profusely to anyone within earshot (which is much further than Clive ever thinks) I would stand, hands on hips, awaiting the pair's return to the farmyard.

'Yer can't yell at Bill like that,' I'd say crossly.

'It's like water off a duck's back,' Clive would mutter. 'Bloody dawg, getting me into trouble.' And he'd scowl at Bill, who would nonchalantly cock his leg on the quad bike wheel, oblivious to the domestic argument.

During those first few months at The Firs it really was a voyage of discovery, getting a feel for the place and trying to get a plan of action together. Outside was comparatively simple really, it was where we felt at ease, we could wall, dig and plant to our hearts' content, but interior design was an entirely different matter. It was the change in the weather that finally drove us back into the house. When all around us became churned with mud, and the rain fell incessantly, then we knew it was time to begin implementing some of the big ideas that we had been talking about since the start of the project.

The onset of bad weather brought one job to the forefront of our minds: the windows. The house had wooden sash windows, and half of them were rotten. Layer upon layer of paint had been applied to the frames but nothing could disguise the water that pooled on the sills inside, and a gale blew between the casement and the panes. The front exterior of the house was cloaked in ivy, tendrils of which had even penetrated through the crumbling mortar. The long, narrow window at the corner of the house, that sat at a right angle to the hearth, would originally have been a fire window, letting light into the inglenook, where the inhabitants would undoubtedly have huddled for warmth on dark winter nights. It is a pleasant thought to imagine the room warmed and shadows cast by a blazing-hot fire, but the reality was more likely to have been a smouldering, lacklustre fire, fed by peat and a little coal of the very poorest quality. When I touched the window frame, it moved, fragments of mortar dropping onto the floor. New fitted windows were going to cost us dearly but until we could make the place watertight and windproof, we could not progress. The old windows were ripped out without much ceremony; the glass was removed by the glaziers whilst the wooden frames and casements were left for us to burn.

The new windows were made to look like traditional wooden sash windows, but were PVC, easier to maintain and considerably lighter on the pocket than the alternative, though not what I really wanted. I was fortunate to have been able to choose: if The Firs had been a listed building then I would not have been allowed this luxury. The difference amounted to many thousands of pounds that I didn't have.

Ravenseat is listed, The Firs is not. It makes no sense that this should be the case: maybe it just got missed, because it is so hidden away. Listings came in during the Second World War, with the aim of ascertaining which buildings would have to be rebuilt exactly as they were, should they be bombed. I imagine that the officials responsible for listing were probably a little more casual about their work high up in the Dales, which was highly unlikely to have been the target of any bombing raid.

'These windows'll keep t'weather and t'noise at bay,' one of the lads said as he admired his handiwork.

'I's not troubled wi' noise,' I said. 'Not 'ere.'

The new windows made a huge difference to the appearance of the outside of the house; the ivy that clung to the walls had been trimmed back and exposed the paint-flecked stone quoins and ledge. A ridge remained where the stone mullions had once divided the windows into two smaller rectangular panes. These would have been knocked out once bigger glass windows became more available and affordable. With the advent of electricity, light is something that we all take for granted but, in the past, it was of far greater value. To live in almost perpetual darkness, through the grindingly long, hard winter months, when the days were short, and the weak winter sun sat low in the sky and the surrounding moors cast a shadow over the farmhouse, must have been dispiriting. However, Ravenseat has a very big arched window built into the east wall, and the morning sun illuminates the hallway and staircase and creeps into the bedrooms, waking us in the most pleasant way. My friend Hannah Hauxwell, the Daleswoman who became famous in the 1970s for living frugally in the most primitive of conditions on her farm in Baldersdale, summed it up perfectly when she said, 'In the summer, I live; in winter, I merely exist.'

Warmth must have been something that evaded the early inhabitants of these houses in the Dales, and centuries later The Firs was still cold. The storage heaters that were dotted around the house took the chill off, the French stove in the sitting room could warm that room itself nicely, but the wood-burner in the larger main living room seemed woefully inadequate for the size of the space. This room had not always been so spacious, as it had at one time been divided by a wooden partition that kept the passageway between the front door and porch to the dairy separate, keeping draughts at bay. The constant footfall of the clog-wearing occupants had worn a shallow groove into the stone flags, and you could follow the dipped pathway where people had rounded the corner into the dairy. Cosiness had had to be sacrificed in order to create space, but now as I sat in my coat, right up against the small wood-burner with my hands nearly touching its cast metal, and I was still cold, I knew that a solution was needed.

'Now this won't do,' I said grumpily to Clive one day whilst

I attempted to warm my numbed fingers. 'I just can't get this thing to throw out any heat at all. And not only that' – I was on a roll now – 'yer can only put the smallest of logs through the door. I've got better things to do than spend my life sawing normal-sized logs into miniature wee ones.'

I have got previous form when it comes to over-ambitiously stoking the fire. Once, long before I met Clive, I half burnt my house door down after embers from a huge, partially burnt log dropped in the draught excluder. I'd forfeited a quiet night in beside the fire when I'd got a better offer that evening and, for safety reasons, I hadn't wanted to leave the enormous knotty log burning in the grate, so I'd lobbed it out into the garden. I was extremely lucky that I didn't raze my cottage to the ground.

Back then I was a contract shepherdess, working on farms all around Cumbria. Once, I had worked on the farm of an elderly bachelor where the farmhouse's living conditions were none too salubrious. On the occasions that my presence was required, usually for mucking out some overfilled barn bottom, I was always curious about the appearance of a dead rat or two beside the house door. One day I decided to ask.

'Yer must 'ave a gay good cat that can catch yer rats o' tha' size,' I said.

'Nae, it's mi lal' terrier that gets 'em,' he replied.

'An' it brings 'em to yer door?' I said quizzically.

'Nah, he gits 'em upstairs an' I drops 'em out o' t'bedroom window,' he replied.

His rat infestation was soon to be unexpectedly solved.

He and I had little in common other than a shared love

for all things burnable. I suspect that he was just too miserly to buy a load of logs, whereas my passion was born of necessity as I really didn't have two pennies to rub together. Tree branches and limbs – lopped off by gales – gateposts and pallets were all gathered up from the fields and conveniently but unceremoniously piled up beside the farmhouse door, at the opposite side to the rodent corpses. As he lived alone and didn't have anyone to tend the house fire whilst he was out in the fields or farmyard, he had devised a cunning way to feed it. One of the many lengthy pieces of timber was suspended at a forty-five-degree angle on a chain from a hook he'd hammered into a beam in the ceiling. The fire would effectively feed itself: as the end burnt away in the grate, the wood would gradually move into the flames.

This basic contraption worked fine for many years, but when it failed it did so in a most spectacular fashion. A fire broke out whilst he was down in the fields on his tractor muckspreading, and smoke billowed through the windows and seeped from underneath the roof slates to such a degree that the rats called time and evacuated the premises.

The fire brigade was called, and the flames quelled, before too much damage was done. Those brave firefighters did their duty and the house was saved, but despite this the old farmer's protestations were loud and clear.

'Thoos put mi fire reet oot,' he squawked. 'Reet out. It'd bin better if thoo'd 'ave left it 'alf in.'

Clive was in agreement that we needed to do something about the wood-burning stove. Radiators are all well and good for background heat, and we would have to install central heating to warm all of the rooms in the house, but nothing could replace the homeliness of an open fire. The hearth is the heart of our home and the focal point for everything at Ravenseat. Babies have been born beside it, beloved dogs have drawn their last breath by it. Boots have been dried there, bread proved, kettles boiled and chestnuts and marshmallows toasted over it. We'd feed the fire lovingly all day with logs of hardwood or sometimes a knotty branch or root that refused to be cut; we were always sparing with the more expensive coal until the evening. The fire is always in, and our house is always warm, and that was what was needed at The Firs.

The antique black range was still in place, a masterpiece of Victorian craftsmanship which took up nearly half of the living-room wall. There was an oven on one side, a water boiler at the other, a drop-down griddle on which to cook, and a brass rail over which towels could be hung. The modern wood-burner had been fitted into the space where the fire grate should have been. With just a little modification, I could see no reason why we couldn't put the grate back.

Above the mantelpiece hung a simple small picture, a woodblock print by writer and artist Marie Hartley, who meticulously wrote about all aspects of the life, tradition and history of the Dales and their folk. Titled *Jack o t'Firs*, it showed a gentleman, long and lithe in a jacket and buttoned gaiters, reclining in a rocking chair beside the fire. A dozing cat is curled by his clogged feet and he smokes a pipe as though in quiet contemplation. Across the fire hangs a crane upon which is suspended a kettle. It was an image that paid homage to a man who had lived and farmed at The Firs nearly a century ago. Susan had insisted that he remained in situ, and all throughout the renovations he never shifted. Dust settled on him and occasionally I'd give him a cursory wipe over with a cloth. Maybe I am overly superstitious, but I liked Jack overseeing things, and that unassuming little picture of contentment conjured up in my mind the idea of how I wanted The Firs to feel. I wanted Jack to be comfortable with what I was doing.

We needed expert advice regarding the fire before we ripped anything out. Perhaps the wood-burner had been installed because the chimney was in a poor state. It looked like a flue pipe had been put up it and the rest of the chimney blocked off with a heatproof sheet. I was loath to ring Susan to ask any questions about it. I didn't want her to feel that we were finding fault with her cherished home or somehow questioning her tastes. We had already agreed that when the renovations were complete, she and Graham would come back for a look and I wanted her to approve of the changes we'd made but, as it stood, with units ripped out and walls with plaster hacked off, it all looked too brutal.

With the gradual disappearance of the open fire and the traditional black ranges, a lot of knowledge on their workings has been lost too. Fortunately, I knew of a retired builder, Alan, who, although still spritely, was not so keen on going up ladders anymore but was happy to come, oversee and impart some knowledge and guidance in an advisory capacity. As a young apprentice he had worked on the traditional black ranges, first repairing and maintaining them and then, as time went on, taking them out and installing a nice electric or gas fire in their place.

I was acutely aware that my nostalgia for the traditional farmhouse cast-iron range-oven was due to the fact that I was no longer entirely reliant on them. The daily work, the sheer drudgery of cooking, heating water for washing and warming the home (all reliant on the range) must have been wearying and there was little wonder that domestic servants and maids were employed even on the smallest of farms. Even the condition and upkeep of the black range itself was considered to be a matter of great pride. There'd be no tarnished brass work or sooty fingerprints, as everything would be highly polished at least once a week, and curlicue decorations were even carefully etched out onto the hearth stone with the use of light-coloured sandstone to add that extra flourish. This artistry was also used on doorsteps to impress visitors to the household. Alan was fond of telling how it was customary for Monday to be cleaning day and any black lead polish that became ingrained in the fingernails would be dislodged on Tuesday, bread-baking day. The convenience of the new electric fires and cooking hobs must have been most welcome: just at the flick of a switch there'd be hot water and a warm house, but without the soot.

Our neighbour and friend Jimmy Alderson grew up at Stone House, the next farm along from The Firs, in the 1920s and '30s. He experienced firsthand how life was before the introduction of modern utilities – or should I say that his mother did. Seeing the black-and-white and sepia photographs of a young Jimmy with his parents and siblings at the farmhouse door always made me wonder whether the stern expressions were a reflection of the hardships endured, or whether it was just 'not the done thing' to smile. It did make *me* smile to find a picture of Jimmy's mother, once again standing outside the door of a house, but this time it was her new-build fifties semi-detached home on a housing estate in Richmond.

'That must have been hell for her!' I'd said to Rachel, her granddaughter.

This was a woman who had brought up her family in the most rural of places, who milked cows every morning by hand, sewed, baked oatcakes, cured hams and worked in the fields.

'Did she not yearn to be back in the hills again?' I said, studying the wiry figure in the photo with her headscarf knotted tightly under her chin and a stony countenance.

'Are you joking?' said Rachel. 'She loved it, she was warm for the first time in her life, had friends next door and could walk to the shops.'

I was pleased that Alan was happy enough with the existing range to give us the go-ahead to pull out the wood-burner. Clive peered up the chimney.

'What can thoo see up there, is there a gurt 'ole?' asked Alan. 'Can yer see t'sky?'

Clive confirmed that there was a bloomin' gurt hole and that he could indeed see the sky.

Next, Clive was instructed to go outside and watch for smoke: hopefully there'd be none seeping out from cracks in the chimney.

We lit a fire beneath the chimney where the grate would be if we had one. It took a few goes to get the paper and

My brood outside The Firs. From left to right: Reuben, Miles, Clemmie, Sidney, Annas, Edith and Violet. Myself, Raven and Clive, who is holding Nancy, are in the background.

Springtime means lambs, and we let our yows
give birth outside as nature intended.

Lambing time can be tiring work . . .

Where lamb and pram go, the mother will follow.
Our packhorse bridge is in the background.

A traditional barn in a wildflower meadow.

The Firs nestled in its valley – it looks like paradise!

When the renovation started I knew we had a project on our hands, but it was daunting at the start.

(*above left*) The children all helped out. Raven and Edith with paintbrushes in hand.

(*above right*) Clive built a seat in the retaining wall. Not the most comfortable place to sit!

(*below*) Bringing the sheep into the pens at clipping time.

(*above*) Our beloved Little Joe having some breakfast with Clemmie.

(*below left*) A different kind of guest in our shepherd's hut.

(*below right*) The occasionally irritable Josie out in the snow with Raven.

(*above left*) Joe, a good dog taken too soon.

(*above right*) Whether it's down to the sheep or Little Joe's bottom, we're constantly fixing our drystone walls.

(*below*) The horses are part of the family and I feel very connected to them.

(*above left*) Eartha and her new calf.

(*above right*) Clemmie gathering up the loose hay. We had
a short window to mow, dry and bale it.

(*below*) Pippen on her patch – she's not one to stray too far these days.

sticks burning but eventually a small fire was crackling. For a while, the smoke billowed then rolled, almost clinging to the brickwork, then it spiralled away behind the metal sooker plate above the grate, designed to channel the fumes up and away.

'We must 'ave chosen a new pope,' Clive joked as he came in.

'Eh?' said Alan, looking perplexed, and then went back to explaining about downdraught, and how I'd need a new chimney pot – tall and narrow to be specific.

'Pulling power is what yer need,' he added.

Clive nodded in agreement.

Then after a bit of measuring and head-scratching he gave me the dimensions for the replacement ironmongery required to restore the range to its former glory.

There was only one place that I knew I could find the items on my list: at Hawes, our nearest market town. Ian, the antique dealer, had a shop, situated halfway up the cobbled one-way street. I would always slow down and stare through the windows as I went past in the Land Rover. If I was on foot, I would invariably find myself lured inside. There was always something that would capture my attention, for Ian had a keen eye and knew what the locals and tourists wanted to see. The shop was full of country farm-house furniture. Nothing twee and twirly, nothing French with spindly legs or painted with Farrow and Ball. Everything was solid and fit for purpose, made of heavy oak and elm – all lovely things but unfortunately also too costly for my pockets. It was Ian's shed, just across from the shop, that interested me. In here, in this fusty, dusty overfilled building, were stacked all manner of wonderful things, though, fortunately, not wonderful enough to be in his shop. It was here that the partially dismantled, slightly broken and beyond-repair stuff sat. In here was where I found my replacement cast-iron panels for the black range, a raised grate and a tidy betty. Heavy as they were, they were still fragile, easily broken if dropped, so Ian and I were careful as we put them in the footwell of the Land Rover. They didn't look much, just rusted fragments of rough iron, but to me they were priceless. I paid Ian, thanked him and assured him that I would be back in touch when I needed more 'things', although I couldn't be more specific at that moment in time as to what those 'things' might be.

Poor Ian, I always told him that I was his best, most faithful customer, but the likelihood is that I was his worst. Always wanting something unusual or trivial and never anything that incurred a large bill.

It filled my heart with joy to see the range complete with a jolly little fire burning in the grate.

'That's what people want to sit against on an evening,' Clive said. 'I think Jack an' 'is cat'd be happy.'

I couldn't have agreed more.

We now had two rooms that had sufficient heating, but the rest remained chilly and reliant on storage heaters, apart from the kitchen, where there stood a solid fuel Rayburn that was used for cooking and heating water. It looked the part – what farmhouse kitchen would be complete without an Aga or Rayburn? – and I'm sure it would have worked well with a competent operator, but that is where the problem lay. They are notoriously difficult to control, especially the solid-fuel versions. Realistically, I needed either a brand-new, modern one that ran off heating oil, or to quit with one all together and install a simple electric cooker and a separate boiler.

After weighing up the pros and cons, it came down to the simple fact that even if we got an all-singing, all-dancing new Rayburn it wouldn't take away the problem of learning how to cook with one. I quite like the simplistic needle gauge on the oven door, with 'simmer, bake, roast', but 'vague' cookery is not everybody's ideal. Paying guests would still need to have a usable cooker with rings and temperature dials. The cost, also, was going to be prohibitive, even second-hand on eBay they were very expensive, so after a consultation with a heating engineer, we opted for a super-efficient boiler system.

All I needed now was somebody competent enough to do the plumbing. Plumbers were a little thin on the ground in our area, we usually managed basic repairs around the farm ourselves, but this type of plumbing was in an entirely different league.

I rang around a few plumbers who advertised in the paper as offering 'no-obligation quotes'. The few that I managed to persuade to come out to me would accompany Clive and myself from room to room, usually shaking their heads, occasionally stopping to scrawl indecipherable notes in a notebook. Some seemed more thorough than others and would climb up and peer through the loft hatches, the rest would just spend their time repeating how difficult it was all going to be. All of them were unanimous in sending us estimates that made me wince and Clive laugh in disbelief.

'And that's just for labour,' I'd say to Clive as I read the quotes outlining the work required. 'An' we've still got all t'fixtures and fittings to buy.'

'I think we should wait – we've got plenty to keep us occupied. Summat'll turn up,' Clive said and, as usual, he was right.

In this golden age of the internet and online communication – even if we do have the world's slowest and most unreliable connection – it was a little out of the ordinary to receive a handwritten letter in the post. It was addressed to Clive and he opened it whilst eating his lunch.

'It's fra' Dick,' he said, between mouthfuls of sandwich.

'Who?' I said and carried on feeding Clemmie spoonfuls of yoghurt.

'Dick, tha' knaws,' he said, sounding irritated. 'Stayed in shepherd's hut in t'summer.'

I dabbed Clemmie's mouth with her bib, awaiting some revelation from Clive.

'Yer knaw, he was 'ere when we 'ad that leak in t'kitchen.'

It literally all came flooding back to me! That day, Clive had got up early and gone downstairs into the kitchen to make me my customary cup of tea – the one that brings me to my senses each morning. It was the summer holidays, the children were all off school, and for once there was no mad rush to be up and about.

'What time's breakfast?' asked Clive as we sat up in bed supping our brews. We would always try to get the children fed before starting the cooking of the full English breakfasts that the majority of guests staying in the shepherd's hut requested.

'Not till nine,' I said.

'What's 'appening today then?' Clive asked innocently.

This was a mistake on his part, as what could have been a leisurely lie-in and then a relaxed breakfast was entirely spoilt by me having an almighty stress about what needed to be done over the course of the day. I reeled off the list: I needed to bake, change the shepherd's hut over ready for new guests that evening, there was laundry to do, washing to bring in off the line, not to mention the fact that we were terribly behind with clipping and should really have been gathering the moor for woollen sheep.

'And then there's everything that needs doing at The Firs . . .' I was getting really worked up now.

'It's no good me just liggin' 'ere in bed, is it?' I said. 'I need to be gittin' on.'

Clive took that as '*We* need to be gittin' on,' and lumbered out of bed.

'No rest for t'wicked,' he muttered as he pulled on his socks, but it fell on deaf ears as I exited the room, ready to face the busy day head on.

I went into the kitchen and didn't like what I saw.

'We've got a flood down 'ere!' I shouted upstairs.

'Yer what?' replied Clive.

Reuben heard me loud and clear and was up and out of his bed like a shot.

Quite how Clive had managed to make tea without noticing the steady trickle of water that leaked through the floorboards above the doorway, I don't know.

'Where's that coming frae?' I said, pointing at the ceiling.

'Bathroom,' said Reubs confidently.

'I know that,' I said. 'I need you to be a bit more specific.'

At this point, Clive appeared on the scene.

'Great, this is all I need,' he said. 'I'm not starting plumbing before I've had mi toast.'

This was the green light that Reuben had been waiting for. Out of the door he went, dodging the waterfall, before reappearing with a toolbox. Clive didn't look up from his toast and marmalade as Reuben brushed passed him wielding a sizeable pair of pipe grips. Then there was a knock at the door, coupled with an enthusiastic, 'Morning!'

I opened the kitchen door to see Dick, my mild-mannered guest in the shepherd's hut, standing there beaming.

'Everything all right?' he asked, looking over my shoulder.

'Yes, fine.' I smiled, surreptitiously blocking the doorway to the kitchen to prevent him coming in and seeing the disaster unfolding.

'I was wondering if I could have a poached egg instead of fried? I don't want to be any trouble . . .'

Just at that moment, there was a thump from above followed by the heavens opening. Clive shot upstairs like greased lightning, hollering to Reuben about the stop tap. You could say that I let it all 'wash over me' because that is what I did. Cold water was pouring through every crack in the boards. I stood there, dripping wet, rivulets of water running down my face. I wiped my hair out of my eyes. Above, an argument was also in full flow.

'Poached eggs, you say?' I maintained my composure. 'That's not a problem, it'll be ready at nine. Now, if you'll excuse me . . .' And with that, I slammed the door shut.

The flow had now petered out to just drips but a sizeable pool of water stood on the kitchen floor.

'Mam, it's the shower pump,' came a little voice from upstairs.

The shower pump was indeed the culprit, and soon it was laid outside on the picnic table in what looked like a thousand pieces. On this occasion, Reuben did not manage to work his magic. The pump was damaged beyond repair, the only solution being to buy a replacement.

Dick got his poached eggs at nine and the matter of the flash flood was never spoken of.

Only when he'd left the hut later that morning, and I'd gone to get it ready for the next incoming guests, did I look at the visitors' book. Sometimes my guests of few spoken words would wax lyrical about their stay using a pen. I wondered whether Dick would write of his experience at Ravenseat. He didn't disappoint. 'Great stay in beautiful surroundings. A good night's sleep and an excellent break-fast,' he'd put, then signed off with 'From Dick the plumber'.

Now, in the kitchen, Clive read the letter back to me. There was the usual stuff: how's the family, the weather and how were we getting on with the house renovations? We had mentioned our project at The Firs to him when chatting one day. Then it got to what he really wanted to know. If we were ever in need of a plumber then he was looking for a break, some time away from city life. We would get our own live-in plumber in return for accommodation and meals.

'No way,' I said; I couldn't believe our luck.

We wrote back, inviting Dick to come and have a look at what the job entailed and to see if he could make some sense of our haphazard methods and wayward ideas. Most importantly, would he be able to cope with living in the shepherd's hut during the winter and eating the meals that I prepared with us as a family?

Dick ticked all the boxes: quiet, unassuming and highly knowledgeable when it came to plumbing. All known sightings of Dick found him wearing cargo trousers with numerous pockets containing the vital tools that every plumber required: notebook, pencil, radiator key and a folded handkerchief. He had a penchant for whistling, and he always appeared to be supremely happy in his work. Well-spoken and achingly polite, he didn't once lose his temper or swear when things didn't go to plan. The children found him interesting, but his politeness would often be tested when Sidney or Annas were at their most irritating, repeatedly asking him 'why?' or just pinching little trinkets that caught their eye. They were like magpies. A radiator valve mysteriously disappeared, last seen in Sidney's pudgy little hand; retractable tape measures were a favourite with Annas. Dick had the patience of a saint. As time wore on, he realized that barricading himself in wherever he was working was the best way to avoid being distracted by these light-fingered little ones.

Every evening I would light Dick's stove in the shepherd's hut; there could not have been a warmer, cosier place to stay. We would eat tea together, catch up with what the older children had been doing at school and discuss the progress being made at The Firs. This was the beginning of our friendship. Over the course of the winter months we battled away with the heating and hot-water system. Dick would attempt to explain the complexities of what he was doing but it was wasted on us.

'Just tell us where tha' wants an 'ole knocking through,' Clive would say.

To be fair, even knocking a hole through the walls was a procedure in itself. The primitive walls were made of thick rough stones and invariably wherever Dick said he wanted a hole there sat a boulder.

''Ow big did 'e want this 'ole to be?' Clive would say as we prised out yet another huge stone from its resting place.

'Not that flamin' big,' I'd say as Sidney scuttled through it on all-fours between two of the bedrooms.

Dick would compile lists as he went along. He seemed to have a never-ending supply of catalogues; he'd fold over relevant pages and circle the items required. I got to know all of my parcel courier drivers on first-name terms as more and more packages arrived with supplies for Dick.

I tried to save money wherever I could and bought nearly all of the furnishings off eBay and Gumtree. The downside of these sites was that I couldn't necessarily buy everything I needed in the right order. I bought a leather sofa and chair long before there was a room anywhere near ready to put them in, and beautiful bed quilts when there were no beds to put them on.

I managed to buy a vintage Royal Doulton sink and WC for the downstairs cloakroom which were better quality than anything in any of the plumbing catalogues. I began scouring reclamation yards for a Belfast pot sink and heavy-cast radiators, but this was where Dick put his brogue-clad foot down.

'If you want cast radiators then it must be new ones,' he said, then went on to explain the troubles that old radiators gave him.

'They're not pressure-tested for modern heating systems and they'd need respraying,' he said. 'It's a false economy.' He was adamant. He produced yet another catalogue, this time one for reproduction cast radiators, and I had to concede that they were indistinguishable from the antique ones.

'Great,' I said. 'How many do I need?'

Dick picked up his notebook, studied it for a moment, then looked skywards whilst he mentally calculated the number, all the while quietly muttering to himself.

'Thirteen,' he said. 'Aye, thirteen.'

'Ha, unlucky for some,' I said.

He nodded. ''T might well be,' he said. 'Have you looked at the price list?'

The cost was going to be astronomical: two of the cast radiators would cost me ten times the price of all thirteen of the standard modern ones. I did sulk for a little while, as that meant we wouldn't be having them, but was soon brought back to reality by Clive's succinct words.

'They're just bloody radiators, Mand.'

Modern radiators it was then. A big order was put in at a supplier in Skipton. Heated towel rails, shower panels, tiles, a water cylinder and tanks all needed to be picked up at a warehouse on an industrial estate. It was a bit of a trail and was going to take an hour and a half just to get there, but it was only to do once and then Dick would have everything he needed.

Also, I had, unbeknown to Clive, made another random purchase from the Gumtree website. A slab of granite that I thought would be useful for something, although I wasn't quite sure for what just yet. The picture in the advert was quite smudgy and the description vague. 'Piece of granite' it said, '£50'. Well, the seller wasn't going to get in any trouble with the Trade Descriptions Act. I emailed the vendor asking for a bit more detail. 'It's as big as a door and it's collection only,' came the reply.

I could see that I was just going to have to take a chance on this one, so arranged to pick up the granite on my way back from Skipton. It was at Dent and I was sure that I had seen a sign to Dent – between Ribblehead and Hawes – on the road back from Skipton. How super-efficient would that be, I thought, to be able to kill two birds with one stone. I'd be taking a trailer for all the plumbing accoutrements anyway, so I'd be able to put the granite in the trailer too. I was far more excited about the stone than I was the radiators.

It was going to take a whole day to get everything done. Clive had offered to look after the little children whilst I was away, as it wouldn't have been a fun day for them just riding the roads, but we needed to feed all the moor sheep first thing in the morning before I set off for Skipton. That way the children would be able to play around the farmyard while Clive fed the cows. Later, Clive might go to The Firs to assist Dick – or hinder him, since he'd be taking the children.

The trip to Skipton went without a hitch but, as predicted, it took the best part of the day. Dusk had fallen as I passed Ribblehead and I was in two minds as to whether I should cancel my detour to Dent.

I stopped just before the signpost and looked at my phone. Unsurprisingly there was no signal, so I had no way to contact the seller to rearrange the collection. Then I told myself to stop being silly and to just stick to the original plan; it wouldn't take me that long anyway, the sign said it was only four miles. The road sign also had a picture of a bridge and stated that it was unsuitable for lorries. It wasn't a road that I was particularly familiar with, but I seemed to have a vague recollection of once taking a drive up there with Clive in the dim and distant past. It was certainly a rural road; as I followed the twists and bends there seemed to be few signs of habitation. I came to a sharp bend in the road and, above me, silhouetted against the darkening winter sky, I could make out the arches of a viaduct. This must be the low bridge that is impassable for lorries, I thought. I carried on, never meeting another vehicle. Occasionally there would be a break in the hedgerows and trees that lined the roadside and I would catch a glimpse of the lonely lights of remote farmsteads in the distance. Then the road seemed to narrow and ahead of me, reflected in my headlights, I saw a sign warning of a hairpin bend. I slowed right down, pulling across to the right side of the road as far as I dared to give the trailer I was towing chance to turn without clipping the drystone wall on the left-hand side. It was a blind corner, so I had no idea that I was about to find myself on a very narrow bridge with a low wall on either side.

Even though I was driving very slowly it soon became evident that I had passed the point of no return. I was now on a bridge that seemed to be exactly the same width as my trailer. I stopped and decided that I needed to get out and investigate just how much room for manoeuvre I had. The fact that I couldn't open the driver's-side door far enough for me to get out told me that I was in trouble. I wound the window down and poked my head out. Looking forwards, things didn't seem too dire. I had a few centimetres of room between my motor and the wall. Looking backwards, though, things weren't too clever: the sidelights of the trailer were touching the wall and, although they cast little light on the situation, they confirmed that I was essentially stuck, wedged between the parapets of the bridge.

I sat for a while, considering my options, and came to the conclusion that there weren't many. Either I attempted to reverse back off the bridge and around the tight corner blindly, in the pitch black, or I carried on going forwards and hoped that I could squeeze through. I secretly prayed that no other cars turned up, thinking I could well do without the humiliation of someone witnessing me in such a ridiculous position. As it stood, if I could extricate myself without too much damage then nobody would ever be any the wiser. I decided that I would have to go forwards, so I put the pickup into four-wheel drive and set off slowly. Things looked promising for a few seconds; I focused on the road ahead, resisting the temptation to glance at my rear-view mirrors. Then I heard the noise of scraping: first plastic on stone then graduating to metal on stone. I felt resistance as the trailer was pinned between the two walls, slowing the momentum of the pickup. The pickup's wheels spun on the loose road chippings, and the smell of rubber permeated the air. I began to have serious doubts as to whether I'd made the right call. I put my foot down hard on the accelerator. The noise was excruciating, something had to be bending and I feared it was my trailer rather than the bridge. Something had to give sooner or later. We moved forwards an almost-imperceptible distance, the pickup engine screaming and the temperature gauge on the dashboard teetering just below danger level. Then, miraculously, like a cork being loosened from a bottle, we shot forwards, and I lifted my foot sharply off the accelerator and turned the pickup ignition off. Time for the engine to cool off whilst I went to examine the damage. Using my phone as a torch I walked back around the trailer to inspect what injuries I'd inflicted upon it and the bridge. I would have been mortified to think that I'd caused some structural damage to a historic bridge but, as it turned out, I needn't have worried. Evidence in the form of great gouges in the roadside brickwork showed that I wasn't the first person, nor likely the last, to have found themselves in this unfortunate position. The mudguards on the trailer had taken a bit of a pummelling; the metal on the bottom side of the trailer had been scraped, revealing untarnished silvery aluminium but, other than that, things didn't seem too terrible.

Feeling relieved to have gotten away relatively unscathed I set off again. I wasn't far from Dent or my destination, and in not so many minutes I was heading up a long concrete road towards where the owner of the piece of granite lived. The stress and trauma of the journey had rather taken the edge off the excitement of my new purchase – all I now wanted to do was go home. The chap selling the granite was helpful enough and it was a good job that I had taken a trailer as it was a thumping great slab and incredibly heavy. We only managed to get it into the trailer, and laid out on the floor, using a combination of 'walking it' on end and then sliding it on cardboard up the trailer ramp. Then it was time to go home but certainly *not* the way I came. This time when I turned the ignition key the pickup didn't start, she fired up on the second go but didn't sound quite right. I had no intention of spending the night in Dent and didn't really want to ring Mark, our trusty mechanic friend, so I reckoned that I should just set off quietly towards home. By the time I got to the bottom of Tailbrig hill, some seven or eight miles from Ravenseat, the pickup dashboard was glowing with a plethora of warning lights – oil, temperature gauge, plus a whole host of other neon symbols that looked mightily important. I rang Clive on my mobile phone, warning him that I might not make it back home and to come looking for me in half an hour if I didn't arrive. We did make it, albeit after a slow and painful journey.

Clive came outside to assess the pickup but, bearing in mind that we were both equally clueless when it came to engines, he drew the same conclusion that I did. It needed to go to Mark's garage and be looked at. It was the next day that we unhitched and unloaded the trailer, at which point the superficial damage became abundantly clear. I had no choice but to come clean about the little incident on the bridge the night before. Clive was suitably unimpressed, and it was of no help that Dick drove an Astra van that was twenty years old, if it was a day, and in pristine condition.

'Look at Dick's motor!' Clive had announced, pointing towards the shining red van. 'Norra mark on it, it's in mint condition.'

Dick, being the peacemaker that he was, tried to diffuse the situation.

'No, no, look, the speaker housing has fallen off,' he said, opening the driver's door to show off this barely noticeable flaw. Then, sensing that he perhaps wasn't helping, Dick went back to methodically sorting through all the boxes that I'd brought, making sure that everything was present and correct, whilst Clive forgot his vexation and cooed over the granite slab. It was predominantly black, flawlessly smooth and flecked with what I assumed was quartz, because outside in the winter light it glittered and sparkled.

'Now that *is* a beautiful piece of stone,' said Clive, running his hand over it. 'Must've set yer back a bit.'

'It was fifty pounds,' I said triumphantly.

Dick came back over to look at my latest purchase.

'That's over a grand's worth of granite you've got there,' he said.

My self-congratulation was to be short-lived, though, as later that same day we got the verdict on the pickup's mechanical problem. The engine was beyond repair, the only option was to buy a replacement and this was going to be costly. A thousand pounds for the new one and then the cost of fitting it. My bargain buy had actually accrued me a whopping great bill at the garage.

'Swings and roundabouts, Mand,' Clive said. 'You win some and you lose some.'

3

And Nancy Makes Nine

The moors above Swaledale are dotted with little stone buildings, many of which were used as both shelters for shepherds and watchpoints for gamekeepers fending off marauding poachers. In the past, animals kept by people were valued in a way that we cannot imagine now. Back when poverty was rife, and the workhouse was the only alternative if prices for wool and meat were poor and money ran out, living such a hand-to-mouth existence meant each animal mattered, and sheep were tended every day. Labour was cheap and thus farmers would jointly employ a shepherd to watch over the flocks. Poaching was a big problem, not just by thieves looking for a way to make money but sometimes by desperate men needing to feed their families. Either way, the punishment for being convicted of such a crime was harsh: transportation for life or even hanging.

These tiny buildings have mostly fallen into disrepair, their small proportions meaning they are of little use these days, and their remote locations making them difficult to access. The watching house at the top of Robert's Seat, one of our heafs, stands solemnly in a wilderness of heathery moorland 1,758 feet above sea level. Now, within its lichen-covered, crumbling walls you will find only nettles and the occasional sheep taking refuge from the brutal winds that rage across this exposed plateau. Cherry Kearton (the great-grandfather of the famous naturalist, also called Cherry Kearton, who David Attenborough recently credited with being his inspiration as a youngster) was, in the mid-nineteenth century, employed as the gamekeeper or 'watcher' as it was then known, and would sit in Robert's Seat watching house, ready to frighten off any would-be poachers. Unfortunately, some kind of altercation occurred with a renowned and fearsome poaching party of miners from Weardale. Cherry was shot in the legs and vowed to avenge his attackers. Patience paid off and one evening Cherry once again spied the miscreants, a group of a dozen or so men, armed with guns and dogs and up to no good on the moors. These men would not only take game but sheep and poultry too, so it was in the best interests of the local farmers, as well as the landowners who employed the gamekeepers, to try and put a stop to their criminal behaviour. Farmers from The Firs, Stonehouse, Hill Top, Birkdale, Hoggarths and Greenses all joined Cherry to ambush the poachers, who had by that point decided to bed down for the night in Charles Rakestraw's – of The Firs – cow house. Cherry and the farmers, each armed with a shotgun, laid in wait all night and, at first light, a scuffle ensued during which the leader of the group was hit over the head with the butt of a gun. One shot was fired and the poachers surrendered and were handed over to the authorities.

The children never tired of hearing these stories and being up by the watching house feeding the sheep on the gloomiest of winter days, when the mist swirls around us, makes the tale all the more vivid.

Poaching or sheep-stealing is a crime that still exists on the statute books. Though not so commonplace now, it is unfortunately still a problem, with sheep rustling having seen a resurgence in recent years. Nobody really knows who is responsible; the general assumption being that the unfortunate animals are slaughtered for meat and enter into the food chain illegally – a thoroughly unpalatable thought. We would read the newspaper stories in the *Farmers Guardian* about farmers (often in more densely populated areas) finding the gates to their fields wide open and their stock gone, tyre tracks in the gateway the only sign left by the thieves.

'Sometimes yer 'ave to be thankful for where yer live,' Clive had said as we talked of these horrible stories. And he was right. You would have to be very determined, and in possession of a talented sheepdog, to steal our animals from the open moors.

Gossip travels like wildfire up the Dales but our location meant that we were usually last in line to receive news, by which time rumours had usually been much embellished. So, we were more than a little sceptical when the jungle drums reported that a farmer in the locality, a highly respected man, had been stealing sheep.

'Yer don't think that it's true,' I'd said as yet another rumour was imparted via the telephone.

'Tales is for tellin',' Clive would say, admonishing anyone

who dared believe that a man who had been a huge name in the sheep world and, as a grazier, had his flock turned out at the adjoining moor, could ever be unmasked as a crook.

As time went on and real facts became known there could be no denying that we, and all the neighbouring farmers, had been duped. The stealing had been done over a number of years and therefore remained undetected. A jar of plastic sheep's-ear tags was found, the only plausible explanation being that the tags had been cut out from their neighbours' sheep. Then when hundreds of sheep were removed to a 'safe house' by the investigating police, it all became rather surreal.

Next came an identity parade of the recovered sheep, held at the local auction mart at Kirkby Stephen. The police issued an open invitation to anyone who thought that they might be missing a sheep – or maybe more – to come along and stake their claim.

The atmosphere at the auction was decidedly odd. We found ourselves looking at pen after pen of sheep, mostly elderly yows with no discernible wool marks, flock marks or heaf marks. They stood forlornly on the bare concrete, a sorry sight indeed. They were not quite neglected, but they looked as though they'd seen better times. Plucked from their moorland homes, taken from their familiar patch of ground and claimed by another farmer, they were now lost souls.

It is almost impossible to explain how the sheep were not of a type (even though they were all Swaledales). There are tiny traits that mean you 'ken' your sheep, that identify one farmer's flock of Swaledales from another. These sheep, the stolen ones, did not match. They were a mish-mash, a variety that could only have come from different flocks.

The auction mart was busy, with folks walking up and down the alleys, some leaning against the iron railings and commenting, 'It's a sad day when it's com' to this I'll tell ya.'

The general consensus was a mixture of disbelief and anger that the trust that has to exist between farmers and shepherds in order for animals to roam freely on open common land had been broken. In the pens people huddled watching a man in overalls and wellies astride of a yow. With one hand beneath her jaw, he lifted her head, tilted her face upwards, and momentarily scrutinized her, then ran his thumb over her bottom row of teeth.

'Broken mouthed,' he said. 'She's as old as the hills she com' off.'

His friend, also in the pen, with his glasses on the end of his nose, now studied the yow. 'Whose hills though, that's the question?' he responded dryly.

Then, stepping forward, he studied her smooth horns.

'Nowt,' he muttered. 'Nowt to see.'

The horn burns that are a standard way of identifying horned sheep breeds had been removed with an angle grinder, a suspicious sign if ever there was one, for why would anyone do that?

People trade in sheep, buying and selling is the aim of the game. Every year we sell a consignment of draught yows, our older breeding sheep, which are bought by other farmers who breed from them for a few more years. People would put their own flock mark on their new purchases but never had I heard of anyone removing any identifying horn burns or ear tags.

We found none of our sheep there, but plenty of other farmers did – particularly those whose land was closer to the accused's farm. The case began to make headlines, the novelty capturing people's imagination. Even the police admitted that never had they had to deal with a stranger – and in many ways more complex – case. They needed the expert knowledge of other shepherds and farmers in order to make sense of the way the heafing system worked – how a flock could remain in one place with no discernible boundaries to contain them and how sheep were shepherded and identified.

We had ourselves been considering whether the practice of ear notching was of real value any more, but this incident reminded us that anything that can aid in the identification of your animals is a good thing. An ear tag with a microchip is all well and good but it's also very easy to remove, whereas an ear piercing or notch remains forever.

The case was referred to crown court and the trial was set for December 2015. It was a real defining moment when the idea that a farmer or shepherd could recognize and confirm ownership of a sheep that had no visible identification was accepted as being plausible. Even so, the case was complicated, and emotions ran high throughout the whole debacle. No one could take any delight in the downfall of someone who was once held in such high regard, and the whole incident threw a wave of suspicion onto farmers. People wondered if the whole sheep-stealing crimewave was down to farmers themselves, a few bad apples that had sullied the reputation of the majority.

'I cannae believe it was one of our own!' was the subdued and incredulous comment made by many.

A guilty verdict was delivered by the jury, after they dismissed the defendant's claim that the sheep had turned up at their farm by accident, without their knowledge. Although it was not clear whether the sheep were stolen deliberately, or just not returned to their rightful owners after being gathered in from the moor, either way it was theft.

We, like many others in the vicinity, would never know how many sheep were taken, and what was lost through the bloodlines and breeding potential could never be valued. Just as devastating was the loss of trust, the breaking of the unwritten rules of the 'shepherds' bible' which had guided sheep farmers for generations.

A three-year prison sentence was imposed on the criminal in January 2016.

It wasn't long after we had read about the verdict that I was getting ready to go out.

'You're gonna have to cut back on t'pies, Mand,' Clive muttered as he tried to zip up my dress.

'It's supposed to fit where it touches,' I retorted huffily before I breathed in. I wouldn't normally have required any assistance getting my clothes on – leggings, skirt and jumper was my usual semi-scruffy ensemble – but in this instance things were rather different. Being blessed with 'having the gift of the gab' as Clive so eloquently put it, I had developed a little sideline in public speaking. Every so often I was invited to be guest speaker at meetings of the Women's Institute, rotary clubs and other such groups, and in order to at least look the part, I'd attempt to tidy myself up. I'd found a simple black dress in a charity shop, nothing too flashy, just knee-length with a zip up the back, but it fitted well and looked smart. It was easy to wear (or at least it was before I piled on the pounds). On this particular evening I was going to a church near Harrogate to talk and, as usual, I was running late. I'd been constantly delayed, my own fault for trying to make things as easy for Clive as possible. I'd always try to make a much more impressive tea than usual and then compile a compre-hensive list of what everyone needed to do before bedtime. A guilt thing, I suppose; I felt like I was leaving Clive in the lurch with eight children to look after and I desperately wanted everything to run smoothly.

'There's some pressure on 'ere,' he said, tugging at the zip, which eventually went up, the seams of the dress stretched to expose the threads that held them together.

There was no denying that the dress was getting a bit tighter. I'd last worn it before Christmas and since then there'd been warming casseroles for tea, home-made bread, and steamed puddings and custard too. Maybe I'd just overdone it; it was food for thought all right.

'Yer lookin' a bit portly yerself,' I pointed out to Clive. 'It's all that fine food that I keep makin' thi, cakes an' buns an' sek like.'

'Aye, well it's buns I was thinking of an' all, a bun in t'oven,' he said, his eyebrows raised.

'No way,' I said, then reeled off all the reasons that showed I could not possibly be in the family way. I didn't feel sick, not even slightly; I was breastfeeding Clemmie, who was now six months old; and most critical of all I was still drinking tea, lots of it. It's an absolute given that I can't stomach tea when there's a baby on its way.

'I'll go to see t'doctor next week,' I said, just to appease Clive, whilst mentally making a note to cut back on portion sizes.

It was still in my mind when I was driving back home after the talk that evening. Once upon a time I'd scoffed at the idea of supermarkets being open for twenty-four hours a day. 'What sort of person would want to do their shopping in the middle of the night?' I'd thought. It would appear that I was that person. I thought I could pick up some essentials: washing powder and nappies, tins of beans, packets of pasta – all excellent winter provisions – and perhaps a few chocolate biscuits for the children by way of an apology for their abandonment. I also picked up a pregnancy test.

First thing the next morning before anyone wakened, I did the test. Just one blue line: negative.

'There, I told yer I weren't,' I said to Clive after the bigger children had gone to school.

Weeks went by, cold wet wintery weeks. The ground was sodden and the days were long and arduous. I took trailer after trailer of fodder beets out for the sheep at the moor. Clemmie, being strong and able to support her own head, had graduated from the papoose on my front to the backpack, so she could come too. I'd drive along on the quad bike in a roughly straight line, picking my way around the gutters and bogs, whilst Annas and Sidney sat upon the mound of grubby beets and threw out a few here and there. I would stop and start, roll a few out onto the ground, and then go a bit further. Only when the trailer was empty could I park up and take the spade to begin chopping them. There was no quick and efficient way of doing this – I'd mulled it over in my head so many times – all that could be done was to walk the whole length of the line, chopping with the spade. Sometimes the sheep would come and begin to gnaw at the now-exposed crisp white flesh inside the fodder beets and this would hearten me to carry on, but to see a thinly made hungering yow turn her nose up and look to me in disappointment was frustrating.

The children had red cheeks and chapped lips, they wore tights and socks and woollen gloves underneath waterproof plastic mittens. In the worst weather, Clive and I would take it in turns to do the outside work. It made for longer days, working from dawn until dusk, but at least the children could play in the farmhouse by the fire if one of us was there. Work took its toll on Clive and me; it was tiring and monotonous, day after day of being out in the elements, our hands sore and blistered from the cold.

'Don't sit so close to the fire,' he'd say as I perched on the fender, steam rising from my coat. 'It'll give yer bad circulation.' He'd go on to cite various old ladies from times past who'd succumbed to chilblains, varicose veins and blotchy, mottled legs that were the colour of corned beef. I couldn't stop him going into graphic detail.

Strangely enough, even after all this physical work, I was still fat, and getting fatter by the day. We decided that I should go to the doctor.

'I'm 'ere 'cos I'm wondering whether I'm pregnant,' I said.

I was suitably embarrassed to be uttering these words, a feeling exaggerated by the flash of incredulity on the doctor's usually impassive face. I fidgeted as he leaned as far back in his chair as he could, and raised his eyes to the ceiling.

'Righty ho,' he said at last, looking right at me, 'what makes you think that you might be?'

'This,' I said, standing up, turning sideways on and pointing at my tummy.

'Hmmm, well we can soon find out,' he said, handing me a specimen bottle.

After what felt like an age – me sitting staring at a full-size medical skeleton that hung in the corner, him staring at the stick that lay on his desk – he delivered his diagnosis.

'Inconclusive,' he declared.

'What do you mean by that?' I asked. 'Surely I'm either pregnant or not.'

He explained that there are many anomalies that can affect a test. My mind was now racing.

'Don't worry,' he said as I worried. 'I'll take some blood and find out what's going on. It could all be hormonal.'

All the way home, I fretted. Hormonal! What did that mean? A phantom pregnancy? Our terrier Pippen had once had one of those. I was mortified. Then, of course, there was the unspeakable: that something was growing inside me that wasn't a baby. Clive was worried when I told him. We talked seriously, but not for long.

'Yer need some fresh air, tek yer mind off things.'

Back to chopping fodder beets.

To his credit, the doctor didn't leave me wondering for

long. After two days of me making a conscious effort to not google the possibilities, he rang.

'I've made you an appointment to see the midwife,' he said.

A week later, I was back in the surgery.

'Not that surprised to see you back,' my midwife said, smiling.

I explained that it was actually all a bit of a surprise to me, as there had been none of the usual symptoms and that, up until very recently, I had assumed that it was middle-age spread.

'How far on do you think I am?' I asked.

'Well, we won't get an exact date until you go for a scan,' she said, 'but we'll have a listen now and see what's what.'

I laid back on the examination bed as she ran her hands, and then the Doppler, over my tummy. I was quiet, listening hard, then, through the crackling interference, we both heard a familiar noise. She smiled.

'There we go, just a guess but the fact that I can hear such a strong heartbeat would tell me that you're around twenty weeks.'

I was flabbergasted; I must admit that at this point I doubted her. Five months!

She fast-tracked me an appointment at the hospital for a scan and I went home to relay the news to Clive that he was going to be a daddy again, and soon.

My midwife was right, I was that far along and then some. Twenty-three weeks said the sonographer. My friend Rachel wasn't surprised at the news of this unexpected pregnancy at all.

84

'Everyone down t'dale's been saying that you're having another one for ages,' she said, as we drank tea at the kitchen table.

'Well yer didn't tell me,' I said.

'It's a permanent rumour really,' she said, and shrugged.

There was no nursery to paint, no pushchair to buy and no new baby vests required; the scene was already set, which is just as well, as our time was taken up with shepherding the sheep and lambing the yows. The sheep took absolute precedence: lambing time comes but once a year and how well we do dictates our whole year. Our visits to the new house were now fleeting and when we did go it just seemed to highlight how much more needed doing. I fairly developed an aversion to the place, as I could only focus on the work required rather than what we had already achieved.

Although only a mile and a half away from the farm, The Firs' position is enviable as it is fractionally lower in altitude than Ravenseat and sheltered from the brutal gales that ravage the upper reaches of Swaledale throughout the winter months. Spring comes earlier here, just by a few days, but it is still heartening to see the land wake up and there's a bite of grass by late March, which means that Miles's little flock of Texdale sheep – which are always the first to lamb – take full advantage of the fields that surround the house. Every day Miles would go and fill their hay rack and take them a bucket of feed. Owing to the over zealous removal of barbed wire by Reuben, by now it was access-all-areas as far as the sheep were concerned. They went in the garden, down the riverbank or in the coal house but would all converge on the lawn as soon as they heard Miles approaching on the quad bike, and then squeeze through the stone stile upon his arrival with their hay and sheep cake.

So, little progress with the house was made over the course of lambing time. For six weeks we knuckled down and farmed. Then, as the weather settled, and the days began to draw out, the daily routines that had occupied our time over the previous winter months were put firmly behind us. After tea, when we would have been feeding, mucking out and bedding up the cows, we were now busying ourselves at The Firs. There were two rooms that were blank canvases, devoid of furniture and carpet, and it was the perfect time to splash a bit of paint about – not literally, but that's exactly what Clive did.

Owing to my pregnancy, Clive made certain dispensations with regards to decorating.

'I don't think yer should be going up the step ladder to paint the ceilings,' he said. 'Thee do t'skirting boards instead.'

Crawling about on my hands and knees was deemed safer.

Decorating is certainly not his forte. Too impatient to just dip the tip of the brush in, he would dunk it right in up to the handle. He was no Michelangelo when it came to painting ceilings, care and attention didn't even come into it. Flecks of paint were flicked at the end of every brushstroke whilst his loaded brush would drip into my hair as I worked below.

I could hardly complain about the standard of work when I myself found the business of decorating so incredibly tedious. I'd cut corners wherever I could, painting over the occasional cobweb and frequently clogging paint into holes in the wall rather than going and getting the Polyfilla.

The longer days and improvement in the weather heralded the return of passing walkers and guests to the farm. In basic terms, we never stopped feeding, bedding up and mucking out, only it wasn't cows, sheep and horses, now it was people. We did afternoon teas for the passing walkers, and breakfasts for the guests coming to stay in the shepherd's hut. It was go, go, go every morning and the kitchen was my domain for an hour. Clive, Sidney and Annas would head out into the farmyard, sometimes little Clemmie would go outside onto the front in the pushchair or, if it was raining, sit in the high chair and sweat it out with me in the hot kitchen.

Singlehandedly, I could easily knock out a batch of scones and a couple of full English breakfasts but then getting the trays, laden with plates, toast, juice, tea and condiments, down to the shepherd's hut was a feat in itself. Sometimes I would stack everything on top of each other and carefully pick my way over the uneven ground, other times I'd enlist Clive to carry a tray. This particular Monday morning, Raven, who was coming up fifteen at the time, had woken up feeling out of sorts and had taken a day off school. I had guests in the shepherd's hut, and was cooking their breakfast, two full Englishes. I had convinced Raven that a bacon sandwich would put her back on the road to recovery, and all I required from her in return was for her to help me carry the trays down to the hut. She could carry the tea, toast and juice; I'd carry the cooked breakfasts. I was well versed in the art of tray carrying and my protruding bump could now be used beneath the tray to stabilize things. It was fair to say that the chances of me dropping everything were near zero! What is it they say about pride going before a fall?

That morning the grass was slightly wet. I had hurriedly put on my wellies and watched as Raven set off carefully towards the wicket gate that let us into the shepherd's hut garth. 'Hang on,' I said, 'I'm coming.' And with that, set off on a journey I've taken a thousand times. Down the slope, through the gate and around the corner where Raven stood waiting for me. Just as I reached her, I slipped and fell backwards. Everything seemed to go in slow motion. The breakfasts were momentarily airborne and, even though I was now flat on my back, sprawled on the grass, I still clutched the tray and made a desperate attempt to catch the food on its descent. The resounding clatter of cutlery brought out our terriers Chalky and Pippen, who stood, noses in the air and ears erect, like vultures scenting a possible feast.

Raven looked shocked. 'Are you all right?' she said.

'Yes,' I whispered. 'Get down, will yer, have the shepherd's hutters seen us?'

I was thinking on my feet, metaphorically speaking, as in reality I was by now on all fours trying to retrieve as much of the breakfast as I could before the terriers got to it.

'No sign of 'em,' she said, popping her head up from behind the small hillock where I was now reloading the tray whilst swatting at Chalky. How they hadn't heard the monumental crash I have no idea, but there was just a chance that I could salvage the bacon and sausages.

'There's grass on 'em, Mam.'

'Back to t'kitchen, we can scrape it off,' I said. 'Have you never heard of the ten-second rule?'

I redid the vital bits, the eggs, mushrooms and tomatoes, while the bacon and sausages went into the warming oven. I sprinkled a little parsley on the mushrooms, a pretty embellishment that also disguised any stray pieces of grass that might give the game away. We went back to the hut where I apologized profusely for the delay in breakfast. Raven looked worriedly at the back of my dress that was crumpled and damp from where I'd lain on the wet grass.

News of the incident got back to Clive and I assured him – and Raven, again – that I was fine. Admittedly, it had been a bit of a shock, but I'd been too preoccupied by the business of the spoilt breakfast to dwell on it. The fall gave me a heavy jolt, and I now had a dull ache in my lower back and pelvis, but falling over is nothing new, just part and parcel of life on the farm.

'Honestly we're both fine,' I said, patting my tummy, and I truly meant it.

'Well, anyways, yer off for a scan this week aren't yer?'

'Aye, Wednesday I am,' I said, slightly surprised that Clive had remembered.

'I was thinking you could go and rescue Brenda at the same time actually.'

I knew why he had remembered now; it was to do with a sheep, as most things around here usually are.

Brenda was originally a nameless sheep, quite thin, a bad thriver, and we had taken her and forty of her equally slim-line friends to our rented field at Teesside. They had spent the winter there enjoying the lusher grass and higher ambient temperatures. At the end of March, it was time for the sheep to return to the bosom of their family at Ravenseat, which is when we realized we were one short. Clive and I argued about this, and I conceded that it was possible that I had miscalculated the numbers and had written the wrong figure on the movement licence. But Arthur, the farmer who owned the field and who kept an eye on them for us, had never questioned the numbers or indeed mentioned anything dying. It was a mystery until a couple of months later when a lady called Brenda rang to say that she had a sheep that she believed was ours. Her farm was at Northallerton and her sheep also took a winter break at Teesside, a few fields away from ours, and when she brought her little flock back home she had gained a sheep.

And so, my appointment at the Friarage hospital at Northallerton for a scan was also to be the sheep's – now nicknamed Brenda – homecoming day.

The one question that I always ask when I go for a scan is: 'Just how big is t'baby?'

I never want to know whether it's a boy or a girl, but I do want to know that it's a healthy size and growing nicely. It was all good news: the baby weighed in at around four pounds, a perfect weight for 34 weeks, and had a good strong heartbeat which was a relief after the fall earlier in the week.

'Everything looks fine, a healthy baby, and you're an experienced mother so there's no need to see you again,' the doctor said.

I went away happily, another two little black-and-white scan pictures tucked into my pocket that I could show Clive and the children when I got home.

Nearly another week went by, the sun shone, and the farm now teemed with new life – birds, flowers and insects – in abundance. Summertime was here, and showtime too. I was invited to go to the Yorkshire Show at Harrogate but decided to decline the offer. Not only was I feeling quite huge, but I also had what I thought was a cunning plan. We had eight hoggs that were fit and ready to be sold at the auction mart. They would undoubtedly command a higher price if there was a lack of sheep going into the ring owing to the other vendors being at the Yorkshire show. Clive thought this was a good idea too and offered to load up the hoggs into the little sheep trailer whilst I got Sidney, Annas and Clemmie into the Land Rover. The windows were wound down as it was a gloriously summery morning. Within seconds Sidney began fidgeting and undid his seatbelt.

'Sit thi' sen still,' I said. 'I won't be a minute.'

I stood back and looked at the trailer, while the hoggs peeped out through the ventilation flaps on the side. First, I checked the trailer electrics, as Clive had been known to forget to connect them to the motor. The only thing that was of any concern was the left-hand side trailer tyre which looked a bit flat.

'Clive,' I shouted. He was already heading back off up the yard.

'Whaaaaat?' he replied.

'It's gotta flat tyre,' I said, pointing at the trailer wheel.

'It's just flat at the bottom,' he retorted.

'Oh, ha bloody ha,' I said. 'That tyre needs some wind in it.'

'You're full o' wind,' he said defiantly. 'It's nowt to worry about, it's just 'cos of where lambs is standin' in't trailer.'

'Where's the airline?' I said.

'It's bust,' he replied. 'Reuben's fault.'

I wasn't best pleased and huffily strapped Sidney into his seat again and set off cautiously. There was no rush to get the hoggs to Hawes auction, as long as they were there by 10 a.m. then that would do. The children chattered away about what they could see out of the window but, of course, the only thing that I could focus on was my rear-view mirror and, in it, the tyre that still seemed to be bulging at the bottom. I got all the way up Banty Hill, a steep climb out of Swaledale with a hairpin bend before the spectacular but treacherous winding road narrowed. At one side there is a perilous near-vertical drop of hundreds of feet, and at the other side there is a steep boulder-strewn hillside and the famous deep sink holes from which the Buttertubs Pass gets its name. It was just when we reached the summit of this climb, as the flat-topped peak of Pen-y-ghent came into view, that all of a sudden there was an almighty bang and the sound of metallic scraping. I hit the brakes and clambered out from behind the steering wheel. The trailer had lurched awkwardly over to one side, the tyre was in shreds and the rim now rested upon the tarmac. The lambs jostled around in the trailer.

'You've gotta flat tyre, Mam,' said Sidney, stating the obvious. 'Do we 'ave a spare un?'

'Nope,' I said, pondering my next move.

There's no phone signal at all, not until you get to Hawes, and we were still five miles from there and the road was deathly quiet, not a vehicle in sight. Unhitching the trailer from the Land Rover, driving to the garage at Hawes, and getting a mechanic to come out and change the wheel seemed like the best plan. But, of course, now that the trailer was askew, there was more weight on the Land Rover tow ball and, try as I might, I just couldn't lift the trailer off.

This made me angry. I decided that I should just keep driving towards Hawes, very slowly, with the punctured wheel on the grass – at least that way it wouldn't make such a dreadful noise. Painfully slowly I went, my hazards flashing, hoping a car would come past and then I could get a message to the garage and ask the auction to send someone with a trailer to pick up the lambs.

There was nobody, just an empty road. My theory about everyone being at the Yorkshire Show was maybe right. I got as far as Simonstone, a hamlet on the outskirts of Hawes, when I spotted an old fence post lying next to a wall. I was trying to use this as a lever underneath the ball hitch when, finally, a Land Rover appeared on the horizon.

I moved to the middle of the road, still holding the fence post. The Land Rover slowed down and stopped alongside mine.

'Mandy? What's ta on wi'?' said Ronnie.

I couldn't have wished for a better knight in shining armour. Ronnie Mecca, a friend of ours, is in his seventies and farms at Calvert Houses, below Muker. He was not renowned for his strength or mechanical skills, but rather for his jovial outlook and his permanent sidekick, Tess, an old, slightly portly sheepdog who was sitting on his passenger seat.

'Awww, don't ask, Ronnie,' I said as I lobbed the fence post back towards the wall.

I went on to explain my predicament.

'Yer shouldn't be trying to lift anything in your state,' he said, squinting though his thick-rimmed glasses. 'I's not a midwife, an' I don't wanna be, not at my time o' life.'

'Ronnie, I just need yer to go an' tell the auction to bring a trailer for mi' sheep. I'll wait 'ere wi' t'kids.'

Once the sheep were out of my trailer, I could do the final mile or so to the garage. The tyre was shredded and the rim now very bent, so I couldn't really damage it any more than I already had.

Ronnie saved the day; he organized the rescue of the sheep, then sold them for me too as I was by now running very late. I bought him his dinner at the auction mart canteen by way of thanks.

'Were a decent trade thi' hogs,' he said between mouthfuls, 'a fiver up on last week.'

I was forty pounds better off than on a normal week, but when I'd asked the garage approximately how much my trailer repair was going to be, I'd been quoted fifty pounds. It had not been a worthwhile trip at all.

'Ow'd it go?' asked Clive when I finally got back into the yard. 'An' where's t'trailer at?'

It was fortunate for him that the journey back from Hawes is both a long and therapeutic one. The big skies and spectacular views down Swaledale had had a calming effect and, by now, my anger had gradually subsided to the point where I could almost see the funny side.

We didn't venture down to The Firs that evening, but the following three nights we went down after tea to varnish the floorboards. The children played outside in the evening sun until dusk came and the midges chased them inside. Only one more week remained before the children were due to break up for the school holidays and, already, they were busy hatching their plans for the summer. We had a busy itinerary; clipping was on the horizon and the house renovation was ongoing. There was so much to do, and how it was all ever going to get done didn't bear thinking about. There was little point in dwelling on it, we just had to soldier on. I went to bed feeling tired but fine, but woke up in the middle of the night with the queerest of sensations. I lay perfectly still whilst I came round, then had the most horrid of notions that I was lying in a puddle.

'Oh, this is not good,' I thought as I totted up exactly how many weeks pregnant I was.

Then things became significantly worse when I sat up and realized that the puddle had not been my waters breaking – it was blood.

I was determined not to panic, I had seen the baby on the scan just days before. I knew that she was four pounds and that at thirty-four weeks she was perfectly viable.

I leant over and tapped Clive on the shoulder.

'I think it's time for t'baby,' I whispered as he stirred.

He yawned.

'Now?' he said.

'Yes, now,' I replied firmly. 'I think that yer should ring for an ambulance right now.'

'Are yer all right?' he said.

''Course I am,' I said, and convincingly, too, I thought. It was most definitely a case of mind over matter, and if Clive wasn't worried then I wasn't worried.

It took forever for the ambulance to arrive; typically it had been at Northallerton when the call came in and had to make its way all the way up the dale to Ravenseat. Once Clive had rung 999, he had not been allowed to put the phone down, even though he'd protested bitterly to the operator that he really couldn't provide a running commentary for over an hour on what I was doing as, frankly, it wasn't that dramatic. The phone sat on the table switched to speakerphone mode whilst Clive studied the *Yorkshire Post*. I was laid out, reclining on the sofa, towels covering the cushions that lay beneath my hips. I was hoping that gravity would help to keep my baby in situ until help arrived.

A paramedic got to us first.

'I's pleased to see thee, I'll tell ya,' Clive said to the green-uniformed man as he showed him the way to the sitting room. 'I can put t'phone down now.' And he went back into the kitchen and thanked the operator.

'I need yer to pack me a bag,' I shouted to Clive.

'I need you to stay still,' said the medic to me. 'Yer bleedin' and I need to put a line in.'

Clive attempting to pack my overnight bag was a perfect distraction to being attached to a drip. Every so often he'd pop his head around the corner of the door for advice on what I needed and where to find it. By the time the bag was packed, the ambulance had pulled up in the farmyard, thankfully with no sirens wailing to waken the household. Much to my consternation they insisted that I was wheeled to the ambulance. Clive chirpily bid me goodbye, the doors slid shut, and I was off to Middlesbrough.

The first ten minutes of the journey were taken up with answering questions and form-filling. I held on to the rail bars on the side of the narrow stretcher bed whilst the ambulance bounced and lurched from side to side on the twisting country roads. A drip bag was suspended above me on one side, and a blood-pressure cuff was strapped around my other arm. To say I was uncomfortable would be an understatement, but more than anything I wanted to hold on to this little early baby for as long as possible. I concentrated hard on being as still as possible, focusing on the wall clock that I could see reflected in the blacked-out window. The closer to the hospital I could get before having the baby, the better it would be, but unfortunately only some twenty minutes into the journey the point came when I could not hold on any longer.

We hadn't even reached Gunnerside when the baby slid quietly and effortlessly into the world. A baby girl, small and perfectly formed, who squinted under the bright lights. We stopped momentarily at the side of the road and the paramedic who had been travelling behind in the first-responder car now joined us. He busied himself checking over the baby whilst I watched wide-eyed. He must have seen the worry on my face.

'She's gonna be fine,' he said. 'We just have to keep her warm.'

I nodded; suddenly I, too, felt very shivery.

'Congratulations, by the way,' he said as he swaddled the baby as best he could, bearing in mind that she was still attached to me via the umbilical cord. And with that, we were on our way again. She was premature, smaller and more delicate than a full-term baby, but she was here and safe, and my sense of panic lessened.

'Can yer ring Clive an' tell him that all's well?' I said.

'Not a problem. I'll let him know, then he can sleep easy,' he said.

Once off the country roads, we raced away while I held my baby close to my chest and closed my eyes, lulled by the crackle of the ambulance radio. I was brought back to my senses when the back doors were opened, and the calm and warmth was replaced with the chill of the early-morning air. A team of midwives and neonatal nurses stood ready and waiting, and extra layers of blankets were piled on top of us as we were wheeled into James Cook University Hospital in Middlesbrough. I blinked under the stark, harsh strip lights that lit the labyrinth of corridors. Finally, we reached the labour ward where we were both thoroughly checked over, the cord was cut, and the baby was weighed and dressed in a tiny vest, bodysuit and knitted hat, whilst the assembled throng of medics talked in hushed voices and studied charts.

'What happens now?' I asked as they carefully laid her in an incubator.

'She's going to have to stay on the Special Care Baby Unit for a while,' said one of the nurses. She assured me that there was nothing wrong, it was all very precautionary, and it was just that my baby was small and a little fragile.

'We are going to put a feeding tube in and you're going to see if you can express some milk for her,' she said. 'She'll be fine, you're an experienced mother, but I can see that it's all been a bit of a shock to the system.'

To be honest, I felt a bit weepy.

'I'll ring my husband and let him know what's going on,' I said as she left.

I had lost all track of time, such a lot had happened, and I was sure that the household would be asleep, so I was surprised when Clive sleepily picked up the phone.

'Hiya love,' he said. 'How are you an' the lal' dote?'

I told him that we were both fine but that we couldn't come home right away, and that he was going to just have to hold the fort for a while.

'I'll email some pictures of her through to Raven, so you can all 'ave a look at her, she's so sweet.'

It made me think of how things had changed over the course of the years. When Reuben had been born very prematurely and had to stay in special care, one of the nurses took a picture of him with a polaroid camera. Then we had scanned it and sent it through to a neighbouring farmer who had a fax machine. The result was a very grainy black-and-white picture on shiny paper, but it was enough to give Clive a peek at his newborn son. That was only twelve years ago.

I spent the next twenty-four hours expressing milk and staring into the little incubator down on the special care unit. I had forgotten just how useless you feel when faced with a sleeping baby wired up to a confusingly complex array of sensors and monitors. I would put my hand through one of the portholes of the perspex box, stroke the back of her tiny hand with my finger and wonder what I should be doing. I'd furtively look around the unit at the other incubators, trying to see what the other parents were doing, looking for reassurance that I wasn't the only one feeling fidgety. I had already decided that, looking at the situation objectively, the most sensible thing for me to do would be to go home, organize some kind of rota with Clive and the children and try and restore some order. I talked to the nurses and a plan began to form. I'd go home and fill some little pots with breastmilk for feeds, since being at home I'd be more relaxed and do a better job. I could also pick up some smaller clothes. I still had Reuben's tiny vests and all-in-one suits in a case in the loft.

I followed the signs directing me to the central atrium of the hospital. Even though it was early, people were milling around the cafe. Looking around, I spotted the travel bureau where I could get a timetable for the shuttle bus that took people from the James Cook hospital at Middlesbrough to its smaller sister hospital, the Friarage at Northallerton. Once there, I could catch another bus to Richmond and then I would be on Clive's radar; he could find his way to me in the Land Rover. I had used this route before when I had been carted off to Middlesbrough after I had Clemmie, although in that instance it had been both me and a baby making our way home.

I can't fault the man on the desk, he tried to be as helpful as possible, but once he'd found out where I lived, 'in t'back o' beyond', he broke the bad news to me. The shuttle bus service had been cut, the only bus route now operating would take me into Middlesbrough town centre. I sighed. The fact that it was a Sunday wasn't making things any easier as the buses were more infrequent too.

''Ang on a minute, what about a taxi,' he said. 'What's yer postcode? I'll get a price.'

He punched in a number on the desk phone, wedged the receiver between his shoulder and ear, and armed himself with a pen.

'*Three hundred quid!*' He grimaced, noted my shaking head, and put the phone down.

He apologized and handed me a pass so that at least when I came back in the Land Rover I could get through the car-park barrier and park for free.

I wandered out of the hospital, across the road and towards a bus shelter where a couple of people stood waiting. I shivered, the day was by no means cold but being in the warmth of the ward had evidently softened me. I didn't have to wait long before a bus arrived. I wasn't even entirely sure how far this journey was going to take me, but every mile nearer to home was a step in the right direction. This was stage one, I was going to Middlesbrough town centre. I googled bus routes on my phone. Where could I get to from Middlesbrough? I couldn't believe my luck when I discovered that there was a Dales Bus, a tourist and sightseeing service, that operated only in the summer on Bank Holidays and Sundays! I could get all the way from Middlesbrough to the Buttertubs Pass, only a fifteen-minute drive from Ravenseat. What luck! I kept scanning the small print, convinced that it seemed too good to be true. The cost? Just £7, not bad for nearly a seventy-mile ride that dropped me at the Buttertubs two and a half hours later. I saw things out of that bus window that I'd never noticed before, but more than anything, I was struck by the stark contrast between the urban and rural landscapes. By now, the colours of the countryside were at their most vivid, lush greenery everywhere, swathes of yellow flowers in the fields under the bluest cloudless skies, and as we got nearer to home, the smell of freshly cut grass filled my nostrils. I closed my eyes, a combination of tiredness and sensory overload. Pangs of guilt gnawed away during this quiet time; whilst the other passengers oohed and aahed at the scenery I just worried about my temporary abandonment of my baby. The other travellers were on a day trip whilst I was literally on a guilt trip.

Clive was waiting for me at the bottom of the Buttertubs, leaning against the Land Rover watching the world go by. I thanked the driver and went and crossed the road to where Clive stood. All around us was a patchwork mosaic of small fields, drystone walls and barns that stretched as far as the eye could see. Clive, in his typically understated manner, put his arm around me.

'You've done well,' he said. 'She'll be back 'ome wi' us in no time.'

The children were pleased to see me, Raven especially as the burden of responsibility for Annas and Clemmie had fallen onto her shoulders. I had only been away for thirty-six hours, but it felt like a lifetime as so much had happened, but at home on the farm everything remained the same. That continuation is something that, in troubled times, can either be a comfort or a frustration. 'Life goes on' is how the saying goes, and this is certainly true on the farm.

Clive's little outing down the dale to pick me up had not served him well. The sight of rows of cut grass drying in the sunshine and the sweet smell of the hay that filled the air had made him impatient to get started haytiming at Ravenseat. By now, every farmer in the dale had a common purpose, to gather up as much hay as possible when the weather allowed it. To be left behind and miss your chance would be unforgivable. Come hell or high water the hay needed getting. I began to form a plan of sorts. Firstly, Edith and Violet were dispatched to the woodshed to find the 'CLOSED' sign and told to customize it a little to make it absolutely one hundred per cent crystal clear that I wasn't open for visitors or business. They returned very pleased with themselves, though, upon inspection of their handiwork – 'Mums had the babby' in crooked capitals – I had to wonder whether it would have the opposite effect.

Next, we discussed baby names and drew up a shortlist, from Rowan (very nice, but far too rock 'n' roll in its entirety: Rowan Owen!) to Sheba. Queen of, I thought. 'Cat food,' said Clive.

Eventually we settled on Nancy Grace.

The older children had another week of school before the summer holidays and, with Clive having to spend his days out in the fields mowing grass on the tractor, I was going to be somewhat limited as to what I could do and where I could go, as I would have Sidney, Annas and Clemmie with me. I certainly couldn't go back to the hospital until there was someone to watch over them.

That week was a hard one. The heat and sunshine were phenomenal, perfect conditions for getting our hay in, but the exhaustion was something else! Our days were spent tramping around the edges of the fields, raking back the loose hay cast out by the hay bob around the perimeter walls of the fields. Occasionally Clive, thoroughly fed up of bouncing around the field on a tractor, would give me a change and I'd do a couple of laps in the driver's seat whilst he sat in the shade of a barn with the children and took a breather. Robert, Clive's eldest son, came to lend a hand when it was time to lead the bales home. For me, though, the days were just so incredibly long. I would wait until the older children got home from school, put their tea on the table and then scrub up in the shower. Then I would set off to Middlesbrough whilst Raven would keep an eye on the little ones, Edith and Violet would play, and Reuben and Miles would head off into the fields to help get the bales in.

Nancy was in hospital for nine days, and during that time I went to see her every other evening, carrying little pots of my milk to feed her. It was a 69-mile trip to Middlesbrough and it took two hours whichever route I tried, and the same coming back. By the time I got there, it seemed that everyone else was leaving. Eight o'clock was the earliest that I could make it. I'd sit, feed Nancy, change her nappy and then set off for home at about 11 p.m., finally getting to bed at 1.30 a.m. There was a parents' room at the hospital that I could have used to sleep in, but I never did. I'd be in the wrong place the next morning, and at least the roads were quiet when I was travelling.

The staff filled in a little book about Nancy, so I was kept up to date with what was happening. They gave me a little piece of knitted cloth, asked me to place it inside my bra and then to bring it with me on the next visit. The idea being that it could be put in the enclosed space of the incubator and Nancy would feel comforted by the smell of her mother and home. I could see this being a good idea and duly wore the cloth as requested, only removing it when I went for a shower before visiting. I left it on the flake (the clothes drier that is suspended from the ceiling in the living room) and then replaced it when I was clean. My good intentions, to make Nancy feel at home, did not meet with the approval of the paediatric nurse who caught a whiff of smoke when doing Nancy's obs.

'Are you a smoker?' she asked as she fished the cloth out and resolutely placed it in one of the paper disposal bags on the side of the cupboard.

One of the disadvantages of having an open fire was that occasionally, when the wind was in a certain direction, puffs of smoke would roll back down the chimney. This was the smell that hung on the cloth comforter, I explained.

I was well aware that the Special Care Baby Unit had to be a sterile, clean environment but with its harsh lighting and white, bright scrubbed surfaces I felt that every time I set foot in there, I polluted the place in some shape or form. I'd wash my hands repeatedly and even wore a clean tubular bandage over my bangles so that I could comply with the 'no visible jewellery' rule. When I mentioned that I'd worn the bangles for almost thirty years and that they wouldn't come off without wire cutters the nurses wanted to know how I managed with lambing sheep.

'I use t'other hand and it's not oft' that we have to intervene

anyways,' I said. 'We don't interfere with a sheep that's lambing unless we absolutely have to, they're better left to get
on with it themselves . . .'

I got nods of approval all round.

It seemed a very long nine days but finally Nancy was
given the all clear. She was feeding well, had put on weight
and was ready to come home and meet the family. I'd asked
my friend Rachel if she could watch the children, so Clive
could come with me on this final run to Middlesbrough. I
woke early that morning, excited at the prospect of the day
ahead. Everyone else was still asleep. Clive had had a hard
few days working from dawn till dusk bringing in the hay
crop and the children were now all officially on their school
holidays so had enjoyed a riotous first weekend off. I tiptoed
downstairs into the kitchen to make myself a cup of tea and
went to fill the kettle. The feeblest stream of water flowed
from the tap, gradually petering out to just a pitiful few
drops. Then it stopped completely. I frowned; this was not
a good start to the day. I looked in the kettle. There was
just about enough to make myself and Clive a brew.

I couldn't figure out why there was no water. The spring
at the moor which supplied our water had never run dry
in living memory, it was summertime so there was no frost
danger, and it definitely couldn't be the old 'frog stuck in
the pipe' trouble that we'd often encountered before the
days of public health water-testing. I had a consultation
with Clive, the fount of all knowledge when it comes to
the water supply. He was baffled . . . then annoyed.

'It'll be summat to do wi' t'water treatment plant,' he
said. 'Nae doubt about it.'

We dressed and went across the yard and opened the dreaded green door, behind which were tanks, timers, pressure gauges and more pipes than you could shake a stick at. We never ventured in here unless there was something wrong and, even when we did go in, we could never fathom anything out.

'Just git the helpline number of t'tank,' said an exasperated Clive.

I rang and left a message that I urgently needed someone to come out to the farm to sort it out. By now, some of the children had surfaced; from upstairs came shouts of 'Mum, I can't brush mi' teeth' or 'Mum, bog won't flush'.

'Yer musn't light t'fire,' said Clive, who went on in great detail about how the back boiler might explode if it heated up and there was no cold-water supply.

I decided that I should ring the hospital and tell them that Clive and I would come and pick Nancy up later in the afternoon, hopefully when everything was back in order.

'I'll be late picking up Nancy,' I explained to the nurse on the phone. 'I've got trouble with mi' waterworks.'

'Well, it sounds like you need to be coming to us straight away,' she replied. 'They can be nasty, those infections.'

I hastily explained and told her that I'd ring them and update them on the situation later.

It was four o'clock when, finally, hand in hand, Clive and I walked out of the hospital with little Nancy tucked into a car seat.

'Hell, she's a lal' un,' said Clive.

For all the pictures I'd taken, none had really put her actual size in perspective. She was dainty and her name, Nancy Grace, suited her to a tee.

'From small acorns mighty oaks grow,' I said, and from the moment we carried her through the old porch into Ravenseat farm, she thrived, never faltering, only going from strength to strength.

4

Smelling a Rat

'By 'eck, tha's got a lot o' firewood there,' Clive said, when he saw my latest purchase – four oak panels, elaborately carved with gothic arches and trefoils. All had come to me via a reclamation yard but, originally, they had been part of a country house in the Lake District.

'What's ta plannin' on doin' wi' them?' he said, but the truth was that I hadn't decided. I knew that at one point the living room at The Firs had been split, with a panelled wall forming a passageway from the front door to the dairy. The same had been done at Ravenseat, basically it was just a simple way of keeping living rooms as snug as possible and keeping draughts at bay. The original Dales wooden partitions were simple affairs, constructed from floorboards, but I was considering replacing the long-lost partition at The Firs with these ornate oak panels.

For the moment, the panels were propped on their side in the adjoining garage at The Firs. I had managed to persuade the dealer to deliver them to me, bearing in mind that my previous outing to pick up the granite worktop had landed me with a large bill from the garage. The dealer had also brought a chandelier that I had enquired about after seeing it on his website. This was no dainty crystal-embellished affair, it was a rustic wrought-iron one which hung from a chain, the type that you could imagine Robin Hood swinging from over a banqueting table. No agreement had been reached on the price, it was negotiable, he said, owing to the fact that it needed rewiring before it would be safe to use. As we struggled to carry the oak panels from the van to the garage, he commented on a small table that had previously stood in the dairy and was now languishing in the garage.

'Nice thing, that,' he said, nodding towards the table, as we took a breather before carrying in yet another heavy panel.

I bit my tongue; I wasn't going to tell him that it was about to be chopped into kindling and fed to the fire.

'Mmm, I'm thinking of painting the legs – shabby chic. Reuben's gonna sand it down for me.'

He went over to the table, pressed his weight down onto it, rocked it and then slid open the drawer underneath.

'It's solid enough,' he said. 'It'll clean up nicely.'

We agreed that in exchange for £150 and the table, I would get the chandelier. That sounded good to me, so we shook hands on it.

Clive was not so enamoured with the chandelier and pointed out that there were not so many rooms in which it could hang, as the beamed ceilings were quite low. What I hadn't yet mentioned to him was that I was thinking of getting rid of the dated plasterboard ceiling in the front porch, and that would be the perfect place for it.

I liked the porch, which was roughly built with an oak door to the left-hand side. It had once been whitewashed both inside and out, though now its outer walls were swathed in a thick covering of ivy. Perched upon the apex of its gabled roof was a stone finial, weathered over time to the point that its original shape and form was now indistinguishable. A simple little window looked eastwards over the footpath from Keld, forewarning the house's occupants of any impending visitors. Unevenly set steps, then stepping stones, guided visitors from the footpath to the porch door. In the wall beside the door, obscured from view by weeds that had pushed through the cracks in the flagstones, was a semi-circular niche that housed an iron boot scraper. Built into the wall, such things would have encouraged would-be guests to remove mud from their clog soles before crossing the threshold. I thanked my lucky stars that wellies were now the footwear of choice around the farm, as I should not have liked to have been standing guard at the door barking at the children to scrape their clogs before coming inside.

'The Firs, a cosy house, older and less altered than most in the dale, and with a whitewashed porch like a dairy; on its window sill a white cat usually dozes.' This was Marie Hartley's description of the house and porch, written nearly eighty years ago, and this picture of homeliness that she paints was what I wanted to achieve. In its present form, the porch was structurally sound on the outside but seriously affected with damp on the inside. It had been used as a tackroom; saddle racks were attached to the walls, as were hooks upon which bridles and head collars could be hung. The porch could still be used for storing tack and boots, but it needed to be opened up and made airy, maybe even have some form of heating incorporated to keep the damp at bay. Mouldy walls were one thing, but mouldering English leather saddlery was a no-no.

When I shared my vision with Clive, he wasn't so suited.

'So yer think tha' we should start knocking ceilings down now, do yer?' he snapped. 'We'll nivver 'ave this place finished if you keep findin' more work to be done.'

'Look, this is what greets people when they come 'ere,' I said, pointing upwards, trying to draw his attention to the water-stained plasterboard ceiling. In an attempt to emphasize my point, I picked up a broom that was leaning against the porch wall and poked at the plasterboard directly above with the handle end. The board disintegrated around the broom handle and years' worth of dust and debris that had accumulated above showered down upon us.

'You've done it now,' Clive said, resignedly.

I hadn't intended to do anything quite so immediate, but the fact that there was now a gaping hole in what was left of the ceiling did force the issue.

Clive went outside to rid his hair of all the loose debris whilst I set to work with my new improvised tool, puncturing the ceiling at irregular intervals and watching as the crumbling panels caved in and crashed to the ground. By the time Clive had dusted himself down, the porch was a picture of devastation. Splintered lats, lumps of plaster, mouse droppings and all manner of rubble lay in heaps on the floor.

'Well, I hope yer like yer new-look porch now,' said Clive.

'I's thinkin' that my chandelier would look well in 'ere,' I said, looking up to the now-exposed roof timbers.

What started out as a rather insignificant little space gradually morphed into something quite grand. The reclaimed oak panels were used to cover the walls. They were not a perfect fit, though, but with the addition of yet more oak panels, this time out of Ian at Hawes's shed of distressed antiquities, we were able to piece together the panels to very pleasing effect. A retired joiner by the name of Ken was called upon to do the refit of the porch. It broke my heart to see a saw being taken to the antique panels but there could have been no better person to do the job. A likeable and jovial chap, he could be entrusted with any task I set him. Years of working with heavy machinery had left him very deaf; I'd often converse with him whilst we were studying the timber and say where to make the cuts, only to find that he hadn't heard a word of it. This, coupled with the fact that he loved singing at the top of his voice, meant our communication was very limited, which is probably why we got on so well.

'Measure twice, cut yance' was his favourite saying, and for this precision I was eternally grateful, as the end result was a joy to behold. The panelling looked as though it had always been there, as it was dark, heavily decorated and had a rich patina that can only come with age.

'It's like goin' into a confessional,' said Clive when, finally, all the panelling was in place.

'Christ, don't start confessing,' I said. 'We'll be 'ere all day.'

The religious theme continued with the purchase of a pew. Being less of a stickler for the precise measurements that Ken was so good at, I just measured out the space in the porch in welly lengths. The auction house at Leyburn had a general household sale once a fortnight, and here you could find all manner of curiosities. The problem was that I would set off with firm resolve, to buy something useful and practical, but then get sidetracked.

'Oh, you'll nivver guess what I found at the auction house,' I said to Clive on my return.

'I'll bet that it weren't what you set out to get,' he retorted.

'A gnu, a stuffed blue gnu. Came out of a castle in Bavaria, apparently.'

'So, you set off this morning with the intention of buying a pew, but instead bought a blue gnu!'

'Pews are ten a penny,' I said. 'Gnus aren't.'

It was a shoulder mount, a splendid specimen dating from the 1930s, and when bidding stalled, I jumped in with the winning bid without really considering just how big the front end of a gnu was. From where I was standing, at the back of the crowd of buyers, in one of the spacious, opulent rooms of the auction house, it had looked rather smaller than it really was. I had no idea where I would display my impulse buy and so, for the time being, the blue gnu (that the children christened Hugh) was draped in a sheet and put in the corner of one of the bedrooms. Hugh did, as time went on, progress from lying on the floor to being laid out, face upwards, shrouded in a dust sheet on one of the beds.

'Christ, it's like a scene from *The Godfather*,' Ken had commented when he caught sight of the recumbent gnu.

I did get a pew at the next sale and, miraculously, it fitted in the porch exactly. All that remained was for the chandelier to be put in place. It was being rewired by an electrician who specialized in antique lighting. Unfortunately, some kind of domestic dispute had resulted in the man's wife confiscating the keys to his workshop and, for a little while, my chandelier was trapped in the middle of some bitter feud. I hardly dared ring to enquire of its progress for fear of stirring up trouble but, eventually, calm and order must have been restored for my chandelier was couriered back, rewired and ready to hang.

Dick and Clive were still running copper heating pipes between rooms, while the children made themselves useful wherever possible. I armed the girls with paintbrushes and set them off to whitewash the dairy. Reuben and Miles set to work on two iron bedsteads that had been left behind by Susan. After dismantling them, they waited for the first fine day and set to work with a roll of masking tape to wrap around the brass parts, and tins of black spray paint to cover the rest of the frames. What a transformation: with new mattresses the beds looked the part, and stylish too.

It was a real test to keep the children amused whilst the monumental task of renovation was being undertaken. Sometimes, on the busiest days when it seemed that we lurched from one disaster to another, I would find myself wishing that I had never set eyes on The Firs. Then, one day, when all of the pipes and radiators were in place and all of the plaster had been hacked off, we saw that we had turned a corner and were on the gradual road to completion. Other specialists were now drafted in to do the jobs that we could not.

Our good friend Django did the electrical work. We had

known Django for many years, he was something of a local personality, an enigmatic figure who was born in Hungary, moved to England, and had a nomadic existence. He had a close affinity with the land and held strong to more straightforward, traditional values in life. He and Clive would debate all that was wrong with the modern world in great depth and had a lot in common.

'Hello, my fellow peasant farrrrrrrmer,' Django would say with his lilting accent, rolling his 'r's.

Django would try to wean Clive onto Hungarian delicacies, winter salamis and sometimes goulash, all laced with paprika and made by his own fair hand out of offal and the bits of meat that most would classify as inedible. He had more success with Bull's Blood, a famous Hungarian red wine.

He had a piebald mare, Gypsy, who was his companion, but most importantly his transport. Though now residing in the dale, he was always a free spirit and rode his horse the length and breadth of Britain. In a western saddle and bridle, panniers loaded with his belongings and with his dog, Lucky, trotting obediently beside him he'd think nothing of setting off for Leeds or some other far-flung place.

Whilst working on the house, he'd chatter away to Dick, recounting tales of his childhood in Hungary and his years of travelling around on horseback. Two more contrasting characters you'd struggle to find, and it amused me greatly to listen in on the mainly one-directional conversation. Django's hair-raising accounts of life on the open road all told in gloriously vivid detail and strewn with colourful language left nothing to the imagination. Dick would listen, nod away in agreement and respond with a suitably shocked 'crikey' at the appropriate moment.

Our builder friend Stephen came to repair, replaster or repoint the walls that required it. And, after my many futile attempts at removing layer upon layer of thick paint from doors with little more than a heat gun and scraper, a chap by the name of Martin came armed with his shotblasting kit. There was a tremendous satisfaction in watching the old paint from the creaking door into the living room being stripped away and seeing the bare wood underneath revealed. Lopsided heavy iron hinges had been attached to equally lopsided battens with horse-shoe nails. A previously invisible repair had been carried out at the bottom of the door; perhaps once it had been a henhouse door for a 'pop hole' had been neatly filled in. Since time immemorial, people had been recycling and using whatever they could lay their hands upon, every object had a story to tell and the door was no exception.

The window seats in the living and sitting rooms were in a poor state. Cracked and brittle, they splintered as soon as they were shotblasted – it appeared that it was only the paint that was holding them together. They were unsalvageable and, once again, Ken was called in to work his magic with the leftover panels from the porch.

As time went on, a display of artefacts accumulated on the mantelpiece beneath the picture of Jack, a collection of insignificant little objects discovered during the works. Yellowing pieces of newspaper that had been stuffed into cracks behind the woodwork, a primitive form of insulation. Fragments of pottery that had found themselves amongst the rubble within the walls, and shards of thick crude glass, clearly of considerable age.

'Yer see folks was battlin' to keep rats out even back then,' said Clive as he studied a lump of lime mortar that was laced with glass.

'Same as they are now,' I muttered.

In the depths of winter, when all was frozen and everything hungering, these old houses became a tempting prospect for the most unpleasant of invaders: rats. Even now, with the advent of rodenticides and sonic deterrents, every winter we would become embroiled in a battle of wits, man versus rodent. Having Dick, the plumber, around, we decided to make full use of his expertise and get him to do a few improvements at Ravenseat farmhouse whilst waiting for Django to do the necessary electrical wirework for the boiler installation at The Firs. The washbasin and WC upstairs in the bathroom needed replacing and, for a little while, we were reliant on using only the downstairs or outside toilet whilst Dick installed the new suite. Many a time I swore under my breath at the inconvenience of it all but would be reminded by Clive that the indoor bathroom and toilet was a relatively modern invention.

We had just been through a hard spell of wintery weather, and the farmhouse had been plagued with mice. I had mouse traps set in the dairy and behind the cupboards and, although annoying, the mice weren't really causing any great harm. Dick had been busy upstairs in the bathroom making new holes in the wall for the waste pipe and had, after tea, gone down to the shepherd's hut for the night. Now, nothing in the world will ever stop a determined rodent from finding its way indoors. They are opportunists and will sneak through a door left ajar in broad daylight, scale a wall or even swim through a U-bend – so a fresh uncovered hole in the wall was just asking for trouble, though we didn't realize it until it was too late.

We had all gone to bed, the household was in total darkness, everyone sleeping, when I had to answer a call of nature. I trundled down the staircase in semi-darkness and then into the bathroom. I heard a rustling noise coming from the farmhouse dairy just along the passageway but didn't think too much of it – it would probably be Pippen on the trail of a mouse. There have been instances, on particularly cold nights, when one or both of the terriers have stayed in the house overnight, although usually the last job on an evening was to open the farmhouse door and see them scamper off into the darkness.

The door to the bathroom was wide open and a low-wattage bulb cast a dim light into the passageway. Half asleep and mid flow I looked to the left towards the dairy just in time to see a rat cornering at high speed and heading towards me. In a matter of milliseconds, the rat launched itself off the single step down into the bathroom, skimming my knees as I leapt to my feet. I didn't see where it went after that, I screamed and hurtled out of the toilet, slamming the door behind me. This was more accidental, rather than a deliberate attempt at imprisoning the wretched creature, as I was in a state of terror. I shot up the stairs, three at a time, switching every light on that I passed whilst screaming blue murder. Reaching our bedroom in about three seconds flat, I pounced on Clive in roughly the same way the rat had on me.

'Whasssup, whassup?' Clive hollered, sitting bolt upright in bed.

'It's a . . . it's a . . . *rat*! An' it jumped on mi knee whilst I was on t'loo.'

'Oh,' said Clive, lying back down again. 'Yer mun't freeten' mi like that, I thought summat reet bad 'ad 'appened.'

I pointed out in the strongest terms possible that in many people's minds what had occurred was 'reet bad'.

'I am *not* going back down and into t'toilet agin until it's gone,' I announced.

The rest of the night I lay awake listening to the gnawing of wood. The rat was trapped in the bathroom and trying to eat a hole in the door. As night gave way to morning, it clearly became more desperate. I listened as the door now shook, the rat rattling the sneck as it tried in vain to make a getaway. We got up early and went to summon Chalky. The children were directed to the outside toilet and warned that under no circumstances should they venture into the downstairs bathroom.

'Don't tell 'em what it is,' I hissed to Clive, although it was obvious to the older ones what was going on, especially when Chalky was propelled into the smallest room and the door hastily slammed behind her.

'Is there a rat in there?' asked Raven with an impassive look on her face.

'Erm, yes,' I whispered, 'but don't tell the lal' uns.'

It took all of a minute for word to get around the kitchen table that there was indeed a rat in the downstairs bathroom. Dick came in for his breakfast.

'There's a wat in there,' said little Sidney, pointing enthusiastically towards the bathroom door. Inside could be heard

Chalky scuffling about, punctuated by an occasional whine and the sound of frantic digging.

'Crikey, a rat,' said Dick flatly, seemingly neither shocked, horrified or even surprised. It was as if for all intents and purposes it was an entirely normal thing to encounter over coffee and toast.

After a short while, Clive became impatient and decided to take matters into his own hands, arming himself with the nearest weapon to hand, a Vileda Supermop. The children ate their breakfast and Dick went upstairs to continue with the work in the bathroom, whilst from the downstairs toilet came the sounds of a room being turned upside-down. Chalky was in her element, yipping excitedly and growling, whereas Clive most definitely wasn't in his element.

All of a sudden, there was a furiously loud flurry of activity. The children stopped eating their cornflakes and stared open-mouthed towards the bathroom door. Finally, there was a dull thump, a squeal (but not from Clive) and the door flew open. Exiting the bathroom wielding a sizeable flattened rodent impaled on a snapped mop handle was a triumphant Clive. Chalky, who followed, seemed to have grown in stature and stood squarely in the hallway, her coat tousled and her chest out. She was the champion and defender of our home and she was proud.

'Diiiick,' bawled Sidney, still sitting at the kitchen table with his spoon poised, ready to resume shovelling the cornflakes into his mouth. 'We got the waaaat.'

Dick popped his head over the landing bannisters. 'Crikey,' he said, only this time he did seem more taken aback.

*

We had learned by now that Dick took most things in his stride, but his patience was about to be tested again when our band of willing helpers at The Firs was swelled with a new recruit we'd actually met the previous summer. When the weather warms up, we have a constant stream of visitors to Ravenseat, many on foot, people out enjoying an afternoon in the Dales or just passing through on the Coast to Coast footpath. Some even made the trip out to us for the sole purpose of having an afternoon tea. One afternoon, after a particularly trying day on the farm, Clive came into the kitchen where I was preparing tea to tell me of a visitor that he'd been talking to. Apparently, they'd had a particularly long conversation about poetry, specifically Robbie Burns, whose works Clive was rather fond of. One of the very first gifts that I ever bought him was a set of books, *The Complete Works of Robert Burns*. I hadn't had Clive down as being in any way poetic, but he could recite many of his poems, his favourite being one written about Burns's yow, Maille.

'He's an interesting chap,' said Clive. 'Lives in a priory.'

'Heck, he must have some kind of addiction,' I said. 'You only check in to there when you've got a serious problem.'

Clive looked confused. 'A priory, not The Priory. He's a monk,' he said.

I admit to being a little intrigued; I'd always been curious about modern-day monks and nuns ever since I visited a stately home as a child and saw Carmelite sisters going about their work in the kitchen gardens.

'So, what's he doin' 'ere then?' I asked.

Clive explained that the monk was on a sabbatical, staying in a local bed and breakfast and spending his days in search of quiet locations in which he could gather his thoughts and paint the landscape. 'Kinda arsing about really,' he summarized.

The next day, Brother Francis, as he was called, was back and, once again, sat on the garden wall talking to Clive about a range of topics. I had a look at his paintings; he was clearly a prolific artist as the folder that he'd brought along containing his work was full. All watercolours and all seemingly executed at a frenetic pace. They were modern in their style, impressionist, I think would be how you'd describe them. Though not particularly to my taste, they were atmospheric and colourful. But I was far more inter-ested in studying Brother Francis, who I guessed to be in his seventies. He had an upright and poised posture and the mannerisms of a learned man. He seemed to be deeply enthralled by whoever he was in conversation with, leaning forward as though hanging off your every word. He fixed you with a watery blue-eyed stare that seemed overly intense for the light-hearted banter that we were engaged in. He was clearly a reader, an artist and a thinker, though he struck me as more of a bohemian type than a man of devout religion.

His life could not have been any more different to that of ours and yet we found common ground over our love of people, animals and nature. I pointed out that if he had come looking for silence and solitude then Ravenseat in the height of summer was perhaps not the right place, but that only a couple of miles down the road was our rural retreat that we were in the middle of renovating and where he'd be able to set up his easel, paint and reflect to his heart's content without interference or disturbance.

'If being down at The Firs doesn't inspire you then nowhere will,' I said.

I wouldn't say that I have an artistic bone in my body, but you couldn't fail to be inspired by such unspoilt natural beauty.

He reappeared the following day to show us his work; his trip to The Firs had clearly motivated him and he had gone all out and created a stunning canvas awash with summer colours and brilliant skies.

'Yer a dab 'and wi' a brush, I'll give yer that,' said Clive.

Praise indeed, I thought. Brother Francis smiled.

'Actually,' he said, 'I want you to have this picture, but I would like something in return.'

Clive nodded, but I could tell he wasn't as keen on the painting now that it came with a price.

'I wonder if you would consider letting me do more painting for you?'

Brother Francis had us both baffled now. I was going to have to tell him, diplomatically of course, that we were not thinking of opening a gallery at any time in the near future. But before either Clive or myself had the chance to make our excuses, he cleared up the confusion.

'Painting the house,' he said. 'Painting the inside of the house.'

The penny hadn't quite dropped, so he persisted, changing his tack.

'Decorating. I'm really good at painting walls, I've got a great eye for detail.'

Our prayers had been answered. We had just found ourselves a decorator. It was agreed that Brother Francis would get in touch after Christmas to arrange when he could come and start. He could give us a week of his time, and all he required in return for this was a bed for the night at The Firs and an evening meal.

'Do you think we've done reet?' Clive asked later, when Brother Francis had left. 'What if he isn't so good at decorating. I can't think that he'll 'ave 'ad a lot of practice.'

'I'm sure that even monasteries will move with the times,' I mused. 'Cells will probably have feature walls.'

Brother Francis was as good as his word and rang as promised. Clive explained that The Firs, although habitable, was certainly nowhere near perfect. There was heating, but little in the way of home comforts or, in fact, any furniture at all. He reassured us that he was accustomed to a simple, monastic life, had just the most basic of needs and was looking forward to his working holiday. The smallest bedroom was the only usable room, so we moved a single bed into there for him. The bed was destined to become firewood after fulfilling this one last task.

'By 'eck it's an uncomfortable bloody bed that,' said Clive as he sat perched on the edge of it.

'It'll be like penance then, won't it,' I said. 'The equivalent of wearing a cilice.'

When Brother Francis arrived, Clive and I took him to The Firs and introduced him to Dick the plumber and Ken the joiner, who were now busy both tiling the kitchen and installing a new shower in a previously empty downstairs room. I showed Brother Francis to his room, where he unpacked his belongings: a sleeping bag, pillow, porridge sachets, a small CD player and a book on the Holocaust.

I talked him through my decorating ideas; there was to be nothing too taxing, just getting coats of emulsion onto the walls, particularly over the replastered areas. I had already decided that I wouldn't ask him to do any ceilings, as I didn't want to over exert him. I had paint at the ready, plenty of it, in neutral shades mainly. In an attempt to be methodical, I set him off painting the first bedroom at the far end of the house. He did a splendid job, his painting was very neat, and his artistic temperament shone through when he carefully painted around the door frame in a contrasting pale hue. Unfortunately, this brushwork did not extend to the skirting boards. He didn't like kneeling, he said, which confused me rather, given his calling.

Any thoughts of my team of volunteers being like one big happy family were soon dispelled when the music began. Dick particularly enjoyed folk and country music, and was secretly rather good at the PopMaster quiz on Radio 2. Ken was a biker, had a Harley Davidson and liked to listen to Led Zeppelin, Jethro Tull and Pink Floyd. Dick and Ken would happily hum away to the soundtracks I put together for them on my phone – a bit of Elvis, some Queen, songs I thought they'd enjoy – and played them through a Bluetooth speaker that Reuben had found under his seat on the school bus.

Brother Francis liked his Gregorian chant music, particularly Enigma, and the Hare Krishna chant. These would be playing loudly on a loop in whichever room he was decorating. Not a word was spoken by either party, but the volume on the competing devices just kept being turned up a notch or two to drown out the other's music. Once Raven cottoned on to this clash of musical tastes, she decided that she would join in with her very own specially selected tunes. She sided with Ken and Dick, playing Guns n' Roses' 'Knockin' on Heaven's Door', 'Livin' on a Prayer' by Bon Jovi and REM's 'Losing My Religion' at maximum volume.

Brother Francis's 'basic' needs soon started to expand. Initially, just a microwave was required for the making of porridge, but soon there was talk of a toaster being needed, and a fridge. Ken had a spare toaster and brought it for Brother Francis to use, but I was steadfast in my refusal to find a fridge as I reckoned that the larder was plenty cold enough to stop the milk from turning sour. The lack of cutlery was mentioned, then the newly installed shower was apparently not angled quite right.

Brother Francis, it seemed, was quite opinionated and began to get a bit stroppy. Relations between him and Dick and Ken worsened as the week went on, after he destroyed the microwave by attempting to soften a bar of foil-wrapped butter in it, spelling the end of any warmed bowls of soup for lunch.

Still, at least the house was getting spruced up and every wall he painted was another off the list. The main bedroom, I had decided, deserved something a little more vibrant colourwise. Opulence was what I was looking for, a rich luxuriant shade that would emanate a warmth and, dare I say it, give a sensual feel to the room.

Clive had already mocked the ridiculous names on the paint tins – 'Gentle Fawn, Toasted Almond, Emerald Isle' – but Brother Francis took an instant dislike to my latest colour choice.

'Crimson Tide!' he said with a look of abject horror on his face. 'You're going to paint it that colour? Why, it'll look like a brothel!'

Quite how he knew the colour schemes of brothels I don't know, but whether he liked it or not I was adamant that the Crimson Tide was going on the walls.

'I think it's gonna need two coats to cover it right,' I said to Brother Francis. 'It'll look great when it's done, a real feature.'

He didn't look at all convinced but set his jaw and began prising the lid off the paint tin with a screwdriver. Before I left Brother Francis unhappily painting, Dick tiling and Ken repairing kitchen cupboards, I warned them that they should watch the weather closely as snow was forecast. This didn't affect Brother Francis, as he was staying at The Firs, but Ken would not be so keen on the prospect of an overnight stay down there with him. Dick, of course, was staying in the shepherd's hut at Ravenseat, so we could always pick him up on the quad bike or tractor if need be.

Clive and I spent most of the day foddering the sheep, and then retreated to inside the farmhouse when the snowflakes got bigger and began to settle. The schools rang to say that the children from the Upper Dale were being sent home early. I'd only just put the phone down when it rang again. This time it was Dick saying that Ken had gone home rather than risk being marooned at The Firs, and Brother Francis had also left, having had enough of painting and being worried about getting snowed in.

'Probably a wise move,' Clive said.

Within minutes, the phone rang again. It was Dick, to tell us that Brother Francis's car had come off the track, just above the middle gate. He'd walked back down to The Firs and now needed towing out. We loaded the children into the pickup and set off to rescue Brother Francis, who had walked back to his stranded car and was awaiting a roadside rescue. There wasn't a huge amount of snow, just enough to make the road slippery, and with Clive and myself pushing, and Brother Francis accelerating gently, the car found some grip and sped off up the track.

'Seemed in a bit of a hurry,' said Clive. 'But time's goin' on so we might as well pick Dick up from the house 'cos his van mightn't travel so well.'

We had hardly got through the door when a wide-eyed Dick appeared clutching a sponge. Normally unflappable, on this occasion he looked uncharacteristically flustered.

'Has Brother Francis told you what he's done?' he spluttered.

'No,' I said. 'What's happened?'

'Paint, he's spilt it.'

'Where?' I asked, mentally conjuring up a picture of how bad this could be.

'Upstairs,' said Dick, still grasping the sponge, which I now noticed was a pinky-red colour.

'*Crimson Tide*,' I said in horror. 'Where has he spilt it?'

'On the cream carpet in the bedroom. I did try to clean some of it up,' said Dick as I held my head in my hands.

'And in the hallway,' added Dick, to make matters worse. 'Well, actually, he kinda walked through it too, so it's sort of everywhere.'

This time it was Clive's turn to say, 'Crikey'.

I went upstairs two steps at a time, Clive and Dick following behind, and we were jointly confronted with a scene of devastation. Rather than a sumptuously decorated bedroom, the image resembled a scene from *The Texas Chain Saw Massacre* with an already drying puddle of red paint just inside the doorway and spatter patterns up the as-yet-unpainted walls.

'Jesus Christ,' mumbled Clive.

Dick said nothing.

'I've heard of paintin' the town red, but this is just summat else,' Clive said, surveying the carnage.

Dick had been working downstairs on the tiling; the first he'd known about it was when he went upstairs, after Francis had made his hasty departure.

I scrubbed and scrubbed and then scrubbed some more and still that carpet looked like someone had tipped a tin of red paint on it. I was annoyed with myself for not having lifted the carpet before the decorating, but more annoyed with Brother Francis for having been so careless. The odd drip or two of paint was one thing, but this was just terrible.

'Not so bloody saintly, was 'e?' Clive said as he went to the sink to fill yet another bucket with warm soapy water. 'Why didn't 'e just own up, confess? I'd 'ave called him an idiot, but I couldn't exactly 'ave sacked him when he was here voluntarily. And he could've at least helped with the attempted clean-up,' he complained.

When the scrubbing was finally done, and the last bucket of pink water had been tipped away, Clive began blotting the carpet with paper towels.

'Come an' look at this,' he shouted as he threw the damp towels in the bin. 'Evidence,' he said, holding the lid of the bin open, 'from the crime scene.'

Peering in, I could see a pair of leather Jesus sandals I recognized as being Brother Francis's indoor shoes of choice. The soles were red, like the designer brand Christian Louboutins, only in this case, it was accidental rather than by design.

Our heaven-sent painter and decorator had left (but not quite without a trace) and from then onwards all painting and decorating duties fell to Ken, who was happy enough to wield a paintbrush and hum away to his own choice of music.

5

Storm in a C Cup

Our original plan had been to have The Firs house up and running in a year's time. That soon fell by the wayside, but we went into 2017 determined to finish and be ready to rent out by the summer. With so much to do to keep the farm and associated businesses running smoothly, I became even more reliant on the older children, namely Raven and Reuben. They were testing times. I would often have to remind myself that this project was for the greater good and would, in years to come, provide stability and security for the whole family. The frustration of trying to be in a thousand different places at once, and seemingly never achieving anything, was difficult to stomach. Nothing seemed to move along quick enough. Every day that I spent working on the farm I felt should have been spent working on the refurbishment of The Firs. Of course, conversely, I would say the same thing when I was at The Firs thinking that I really should have been at Ravenseat.

At least we weren't living at The Firs, and could shut the door and walk away from the mess while renovations were going on. But then Ravenseat itself started to resemble a building site. The problem started in 2016 when years of torrential rain eventually took their toll and water poured in though the gable end, seeping down the interior walls and forming puddles on the flagged floor inside.

'I'm sick o' this house smelling of cat pee,' I complained as I wrung out yet another sodden rug that, even though it had been inside and by the fireplace, had become saturated overnight during a downpour. 'It's making such a lot o' work for me.'

'I'll get mi mate Stephen the builder to come an' clart some cement on,' Clive said, 'but I'll 'ave to tell t'estate first.'

It was just a formality that as tenants we should alert the landlord of any work we were going to undertake on the property. In fact, they'd often be willing to pay if it was deemed to be a necessity. Of course, after speaking with the agent, what started out as a relatively simple repair job turned into a mammoth undertaking, with listed-building experts, architects and planning officers all getting involved. Clive was becoming increasingly annoyed, especially when the conclusion was that before any work could start there needed to be an investigation into the cause of the leak. We'd have to liaise with the relevant authorities and have an assessment to diagnose the problem before we began to rectify it and find a solution.

He was less than amused when the scaffolders arrived in November with the express instruction to scaffold the north-facing gable end.

'Wrong end o' t'house mate,' he said flatly, whilst leaning

against the Land Rover and watching the poles being erected beside the kitchen door.

The scaffolders begged to differ. 'Look 'ere mate, this is what we were told to do, it says it right 'ere,' said one of them as they waved a piece of paper around.

Clive shrugged. 'All I'll say is that watter is pissin' in at t'other side o' t'house.'

It was indeed the wrong side of the house and, a week later, they returned and scaffolded the correct side. Nobody said anything.

Finally, a decision was made: the roof was to come off and the whole house had to be repointed with lime mortar. There would of course be considerable disruption during this process but the reward, a watertight weatherproof house, would justify the turmoil, we were assured. Clive was not convinced.

'Whaaat the hell, I cannot believe it,' he complained. 'Lime mortar, it'll nivver hod man.'

The architect disagreed and explained that it was how older buildings had always been held together.

'Lang ago they only used lime mortar because they 'adn't invented cement,' Clive complained. 'Square wheels were deemed all right until someone invented round uns.'

I just laughed.

The whole renovation took nearly six months, and for the entirety we tolerated falling masonry, the sound of drilling and the lack of privacy that being besieged by an army of builders entailed. I spent my days apologizing to tourists about the unsightly works and spent the evenings trying to discourage the children from using the scaffolding as a supersized climbing frame. Reuben would do chin-ups by the back door, Edith learnt a few pole-dancing moves and Sidney could exit the upper floor via the bedroom window then slither to the ground down the equivalent of a fireman's pole. I soon found that I, too, could make good use of it by suspending larger items of washing from the metal framework.

Reuben was excited when a big orange excavator rumbled into the yard, having made friends with the driver earlier the previous year. Both The Firs and Ravenseat are situated on the very fringes of the cultivated land of Upper Swaledale, and beyond these outlying farmsteads lie only vast tracts of open moorland that stretch away into the distance. These deserted places, where one can walk for mile upon mile without seeing any sign of habitation or another living soul, are not only rich in biodiversity, home to many rare plants and birds, but are important on a global scale, thanks to the vegetation's ability to lock up carbon. The rivers are the life-blood of the lowlands, providing water for the towns and cities. The moorland habitat's ability to hold water also substantially slows down the flow of water and thus reduces the flood risks downstream in the lowland areas and, as part of a new flood-management policy, the grips (drainage ditches) that were dug post-war to drain and dry out the ground are now being filled in. Slowly but surely, with the use of a tractor and a spade, the deep narrow drains are being blocked up, in a move designed to make the moors wetter and to encourage peat regeneration.

After a particularly wet winter when flooding downstream had become a major headache it was decided that it was time to bring in experts who could fill in the watercourses far quicker and more efficiently than we ever could. Our forty-year-old digger was just not up to the job so a contractor was found, who had the specialist equipment required, and within just a week he had done work that would probably have taken us a month. He parked the large modern excavator beside the moor gate and promised to return in a few days with a low loader to take it away. Months passed, and the bright-orange excavator remained, a blot on the landscape. Whenever I took a picture of the surrounding moorland my eye was drawn to the lurid machine abandoned amongst the heather. The children spent many happy hours pretending to drive it, pressing all the buttons and turning the dials, and gradually the cab of the excavator filled with empty sweetie wrappers. Even Clive and I managed to both squeeze inside once, taking refuge from a snowstorm that was raging outside. The ignition keys were hung from a nail in a beam in the kitchen, awaiting the owner's return, and sure enough one beautiful spring day he meandered back into the farmyard.

'Can I come wi' yer?' Reuben pleaded, I daresay angling for a spell in the driver's seat.

'Why aye,' came the reply from the beanie-hat-wearing digger driver.

The orange digger slowly rumbled into sight, making its way steadily downhill and into the sheep pens. The clatter of metal became louder as the machine jolted and bounced its way home, the tracks scraped and creaked as they navigated the solid ground rather than the spongey grass. Reuben had been given the opportunity to drive part of the way down the fields but had been ousted from the driving seat with the trickier pens now to negotiate. His face was beaming as he ran towards me to relay how they'd turned the key and the machine had started at the first touch.

'You'll nivver guess what an' all, Mam!' he said excitedly. 'There was a nest in t'exhaust pipe.'

Fortunately, there'd been nobody at home when the engine was fired up and the nest was ejected at high speed in a cloud of smoke.

I walked across towards the advancing digger and gave its happy driver a thumb-up.

'I nivver thought it'd start after standing so long,' I shouted.

'Why aye, like, I cannae believe tha' it's gan sa well, like, eh,' he said in his thick Geordie accent.

At that precise moment, Reuben let out a holler and I saw him frantically pointing skywards.

'Jesus Christ – stop!' I shouted at the driver, waving my arms to get his attention.

He looked up and immediately saw that the bucket attached to the extended front boom was not just touching the electricity wire that ran between the farmhouse and buildings but exerting such a pressure on it that the normally slightly sagging line was now very tightly stretched.

The driver swore, and Reuben covered his eyes, waiting for either the cable to snap or, worse still, the digger to be lit up with some kind of electrical surge.

Backwards the driver went and, when he was no longer in contact with the wire, lowered the bucket.

'That were lucky,' I commented when, finally, he came to a standstill in the farmyard.

'Aye, I spoke too soon, man,' he said as he pulled a cigarette from the back of his ear, lit it with a shaking hand and took a long draw.

That was a lesson learnt for Reuben: never be complacent and always be on the lookout for hazards. Being a believer in the 'nothing ventured, nothing gained' approach, I would encourage the children to sometimes take measurable risks, to challenge themselves. In my mind, to know your limitations and to be able to assess the danger and deal with it was a good life lesson.

The excavator had now returned to Ravenseat, in all its orange glory, only this time it was to dig out around the foundations of the farmhouse where the stonework needed repointing.

The spring of 2017 was almost perfect in terms of weather for lambing. Little in the way of rain meant that water levels were low, and the becks ran almost dry. Hardly a year went by when we didn't lose newborn lambs through drowning so not having this danger to contend with was wonderful. Each and every day dawned dry and bright, not so warm as to bring about the onset of ailments associated with excessive heat – rattle belly and scour – and not so cold that we'd be troubled with pneumonic lambs or the yows with mastitis – known also as blackbag, a descriptive name taken from the symptoms of this dreadful affliction.

Edith is a real asset at lambing time and kens the sheep almost as well as we do. It really is all about thinking on your feet and practical problem-solving. She helped me move one yow with only a single lamb to just outside the kitchen window. The yow was producing too much milk, and Edith and I would take a couple of small jars and fill them with the excess, leaving plenty for her own lamb but producing enough to provide a feed for another lamb whose mother had pneumonia and was very poorly. We didn't have the heart to take the lamb away from the sick yow as she cared for it and nurtured it so well; we were sure that it was the lamb that gave her the will to live. This was all well and good but, like man, lamb cannot live off love alone and its mother, due to her illness, didn't have a drop of milk. Edith knew exactly what the job entailed and could do it all perfectly by herself. The hungry lamb would run to the kitchen door to be fed, coming right up to Annas and Clemmie to be stroked. The pride Edith took in being able to do this was priceless, and the day that the poorly yow was well enough to go back out into the field with her lamb in tow was a cause for celebration.

As usual, in spring we had pet lambs. The numbers vary from year to year, and this time we only had four who were being fed from the bottle. They did very well, roamed the farmyard, and were healthy and happy. They weren't afraid of people and would saunter around the picnic benches greeting the walkers and visitors who came for cream teas. But they didn't like rain, and I spent a fair bit of time shooing them out of the old chapel building where visitors go to have tea in bad weather. If I shut the door to keep them out, they'd loiter around the farmhouse door hoping to spy the children, with whom they were familiar and who they associated with a milk feed. They were not averse to sneaking into the house when I wasn't looking, and it was never only one that ventured over the threshold as they were always together: if you saw one, you saw all four.

After dinner one day I put Clemmie down for a nap, but she toddled downstairs a few minutes later.

'Clemmie, you should be in bed,' I said. 'It's sleepy time.'

'Sheeeeeeepy time,' she said, pointing back up the stairs.

And there they were, all four of them, on the landing. I shooed them back down. Counting sheep was supposed to help you nod off, but this was ridiculous.

There can be no telling when things are going to go wrong, and fate has a nasty way of creeping up on you and kicking you on the backside just when you are not expecting it. After weeks of lambing, it feels like the hard-won battle is over when you finally take the marked-up yows and lambs back to the moor. The lambs, now two or three weeks old, are strong enough to walk, following their mothers the mile or so uphill through the bracken to the moor gate. Ambling quietly behind the flock so not to tire them, with the sheep-dogs weaving back and forth, occasionally nipping at the heels of a belligerent yow when it deviates too far from the course, is a thoroughly pleasant and satisfying task. Your job as a shepherd is not over, but the hardest part is behind you, the yows birthed safely, the lambs past the first few critical days and now the little family units are ready to go forth onto their heafs. For the yows this is a familiar place, where they and their ancestors have roamed for centuries and they can settle, but for the lambs this is all new and about learning the lie of the land, developing the instincts that will keep them roaming but only within the confines of this boundary-less ground.

The shock discovery of the fate of nearly a dozen yows and lambs didn't occur until one fine sunny evening in June. A distant fence line that divided the peat haggs of Cumbria and Yorkshire had become tired and the wire sagged to such an extent that sheep were getting onto an area set aside for peat regeneration. The rarest of bog plants grew in this area and it was often visited by keen botanists who would marvel at what seemed to the untrained eye to be the most unremarkable of flora.

We took part in environmental stewardship schemes, for which we were paid, which meant that some of our land – as was the case here – needed to be clear of stock at certain dates in the calendar to encourage regrowth and regeneration. Clearly, a new stretch of fence was required. New posts, strainers and wire needed to be taken across the roughest of terrains and laid out in preparation for a day when we felt strong enough to begin the laborious process of knocking in the new posts. A tractor-mounted hydraulic fence-post knocker that takes the sweat out of the job would have been welcome but, owing to the remote location and treacherous route through peat bogs and gutters, we had no choice but to opt for man power. Jonny, a local lad and the son of my friend Rachel, was drafted in to help with knocking the posts into the ground.

It took a good few trips back and forth with equipment, as we could only carry a certain amount on a quad bike and trailer. Overload them and you run the risk of getting stuck, owing to excessive weight, plus the route was so rutted that everything would bounce out of the trailer and you'd be constantly stopping to retrieve the fence posts. We took turns to carry the posts and wire over the last few hundred yards, it was just too risky to go any further as the drier ground frequently crumbled beneath us and gave way to blanket bog.

It was on the return journey of one of these trips that Clive and I accidentally deviated from our usual route. The sun was setting, and we were not in so much of a rush to be back to the farm, so we stopped the bike and felt the last rays of sun upon our faces. A light haze hung over the distant heather-clad moors and just the faintest wisps of cloud tarnished the near-perfect sky. It was still, not the slightest breeze to stir the woolly heads of cotton grass that now were bathed in a warm, rich amber light. The silence was only broken by the occasional bleat of a lamb or the almost mystical call of a drumming snipe, invisible to the naked eye but making their presence felt. I stood, breathing deeply.

'Isn't this the most glorious place?' I said to Clive, who was still sitting on the bike. 'It's at moments like this that I feel so in touch with nature and the earth.'

I was working up to saying something deep and meaningful.

'It's impossible to be in such a stunningly beautiful place and not feel some kind of deep emotion, isn't it?'

I closed my eyes, lost in what was almost a spiritual moment.

'Summat stinks,' proclaimed Clive.

I opened my eyes.

'Can ta nut smell it?' he said, now looking this way and that, his nose in the air.

I sniffed and thought yes, maybe there was a smell, a

familiar smell. The stench of death. We both looked around and our eyes were drawn by a black patch of ground only fifty yards or so from where we were. As we walked purposefully towards it, the smell became gradually more pungent until it was almost unbearable.

I stopped, put my hand over my nose, and breathed through my fingers. Clive walked to the black patch and then stood, looking downwards, his hands on his hips.

'What is it?' I shouted.

He turned and grimaced, the colour drained from his face.

'Jesus, yer need to see this,' he said.

What a sight! Never, ever had I seen anything quite like it. The black patch, only some 12 feet in diameter, was thick, oily mud. Ridiculously, it had wavy edges and looked as though a child had drawn a cartoon puddle. There was no real graduation between solid ground and this puddle. It was just there, a surreal mass of black gloop that seemed to sit upon the grass.

Bogs are something that both Clive and I are very familiar with, but never before had we seen anything like this. Usually a bog would be covered with a layer of sphagnum moss in the brightest and most lurid shade of green. The more obvious ones were no secret and would resemble a small pond, standing water on the top, seaves and sphagnum moss growing around the edges.

This bog seemed to be slightly raised from the ground because the surface was bumpy. Upon closer inspection, these bumps became identifiable as sheep. The big humps were yows, and the smaller ones lambs. Clive and I stood, aghast at the horrendous sight that confronted us.

'There nivver was a bog here,' Clive said, shaking his head.

It did look as though it had just burst through the surface, rather like an Icelandic mud geyser that I had once seen on the television, but quite why this natural phenomenon should have occurred at Ravenseat is a mystery. The real puzzle, though, in our eyes, was why had the sheep gone into it? The fact that there was no water took away the theory that it was thirst that had lured them to their deaths. We walked around the bog and saw that in places the outlines of horns could be seen. The profile of the head of a young lamb was just beneath the crust. It was sickening and puzzling.

'Yer'd wonder why, yer really would,' mused Clive.

'Why would so many animals venture into such a tiny bog, when they have thousands of acres to roam?' I said.

'Nay, we'll nivver knaw,' said Clive. 'But we'll 'ave to fence this off now to mek sure nowt ever ga's in it again.'

There was nothing more that could be done other than to construct a fence around the hell hole and leave nature to do its work and reclaim the bodies and, for a while, the bog became an object of fascination for the children. When we eventually set off on the backbreaking task of renewing the original broken fence, the accompanying children would ask to stop just to peer at the bog. There was talk of spectral ghost sheep, Violet and Edith being fond of making up such tales just to see the reaction of the little ones who hung off their every word. For a while, the children were afraid of being out on the moor in unfamiliar places, always watching where they stepped in case they happened upon the same fate. Getting bogged was a frequent enough occurrence, you didn't have to venture far before you could find yourself up to your knees in mud, but seeing and smelling the rotting corpses served as a timely reminder to the children that there is danger to be found even in familiar places.

There is nothing quite like the thrill of a ghost story or the fairy tales, myths and legends that have been passed down from generation to generation, rooted in places that are real. I can't imagine how many times I've had to recount the story of the robber trying to gain entry to the Crook Seal inn, now just a solitary barn standing at the roadside on the way to Kirkby Stephen, how shots were fired through the roof by the terrified innkeeper and then years later a skeleton discovered in the peat. There's usually an element of truth to these stories, even if the facts have been distorted over time, and there's no doubt these folk tales keep the landscape alive and exciting.

Every year we would welcome the return of the house martins that nest in the corners of the sash windows. They had already begun the construction of their new nests when the scaffolding appeared, but they did not seem troubled by the close proximity of the builders, they just carried on flitting back and forth carrying their nest-building materials and, later, food in the form of insects for their broods of chicks. We didn't open the upper floor windows during their breeding season in case the nest was dislodged and crashed to the ground, so a scaffolding platform was a bonus as far as the children were concerned, enabling them to get close and observe the previously secret world of the feather-lined mud nests. Inside, scrawny ugly featherless chicks lay quiet in semi-darkness. Only when an adult bird returned did they come to life, craning their necks upwards with gaping yellow beaks that begged to be fed.

Sidney began collecting insects for the chicks, picking flies from the gauzelike webs in the rafters of the farm buildings and moths from the bathroom window sill. He spent many happy hours filling little pots with dead insects which he would then put on the scaffolding walk-board nearest to the nest. I cannot categorically say whether the insects were eaten or not, but Sidney felt like he was doing his bit.

One morning in July, before school, Sidney marched into the kitchen, the bearer of bad news. The night before had been exceptionally windy, the plastic sheeting that temporarily covered the now partially open roof had flapped noisily all night and there had also been an unseasonably torrential downpour. A summer storm of this intensity was often the death knell for the occupants of the nests, the dried mud slowly disintegrating until finally everything fell to the ground.

'Mam, it's really bad,' Sidney said.

'What's up, Diddler?' I asked, looking up from preparing lunchboxes.

'Mi nest in Raven's window is gonna fall off,' he said sadly. 'It's 'angin off.'

'Ah, now that is a shame,' I said.

'Nature can be cruel,' said Reuben, an unusually insightful comment from him, I thought.

'Bummer,' said Raven. 'I liked to hear the chicks chirruping when I was in mi room.'

Sidney announced that he was going to go and have

another look; I could see that I was going to get a running report and be spared no details. While Reuben and Raven set off to school, Miles went with Sidney to investigate. The builders showed up bright and breezy and were hijacked by a wide-eyed Sidney before they had an opportunity to see whether there had been any overnight storm damage to the building project.

Then the school bus arrived for the little ones. Edith and Violet were ready and waiting, but Sidney and Miles were a-wantin'.

'Will yer come down please?' I shouted up to the gathered throng of hard-hatted and otherwise who had all congregated on the scaffolding platform outside Raven's bedroom window.

'We've mended the nest!' shouted a triumphant Sidney.

He stepped back to admire his handiwork while the builders grinned.

'We scaffold da 'ouse an' we scaffold da bird 'ouse,' the foreman, Julius, said in his thick Lithuanian accent.

Apparently the ever-innovative Miles had spied my under-wired bra, which I'd inadvertently left hanging on the washing line overnight, and, aided by the builders, now had the fragile nest cradled in one of the cups and held in place with the shoulder straps looped through the window.

'The chicks are all right,' shouted Miles. 'There's three, an' they're still in there.'

'Bra-vo!' I shouted. 'Well done, Diddler an' Miles, but it's time for school now, an' time for tea, don't ya think?' I added, turning to the builders.

The builders smiled and nodded. Even Clive had to admire

the handiwork. We stood drinking tea and watched the house martins fly back and forth unperturbed by the addition of extra padding to their nest.

'I think, mi dear, 'twas all a storm in a C cup,' Clive mused.

An unseasonably wet summer is the worst thing for a farmer, and far more difficult to contend with than a hard winter. Summertime is all about preparation for the forthcoming winter and thus we rely on good weather for haymaking, bringing in the crop that will keep the animals fed, fit and healthy over the winter. It rained solidly and continuously through July and August 2017, and when, finally, there was a break in the weather we had only a very short window of time to mow, dry and bale the hay. The ground was wet, the hay wasn't hay – it was grass and it was damp. It was a disaster from start to end but we were not alone, everyone struggled. Spirits were at a low ebb. There'd been plenty of years when the weather had been 'catchy', with no long stretch of dry days, so the haymaking would have to be conducted in fits and starts, but there had been no such washout of a British summer since 1985. Our aim is to get the grass cut, to make hay, before the end of July but if we have to wait until August then so be it, we can cope. But every day of the summer of 2017, when we looked out, it was raining, or the grass was still sodden from the previous day's downpour. It needs to be dry to cut for hay. If we are forced to cut it when conditions are damp then it becomes silage, wrapped up in plastic and preserved by fermentation. This is costly to make, leaving us with unwieldy bales that are impossible to get out to the moor and to the sheep. More importantly, the sheep don't like it as much as hay and if the bale of silage has soil within it then you run the risk of your flock being affected by listeriosis. In most cases this is fatal.

Nowadays we do have another option open to us, buying in hay from other parts of the country that have fared better weatherwise than us. It's expensive and the quality varies. Our own hay comes from unimproved traditional herb-rich meadows that are now a rarity but in buying in crop we buy in security. We can also supplement the forage with extra feedstuff, sugar beet pellets, molasses, fodder beets and sheep nuts – our livelihood depends upon keeping our flock in good order.

Another consequence of a bad summer and late haymaking is that we don't get the fog, which is what we call the new growth of green, very nutritious grass that springs up after mowing. This is perfect for us to put our lambs on when they have been speaned (weaned) from their mothers, the rich grass replacing the nutrition they were previously getting from their mothers' milk. The knock-on effects of the wet weather were astounding, and it's fair to say that there was a distinct lack of happy banter around the kitchen table. It was doom and gloom when we met with other farmers – all with stories to tell of mouldy hay, balers bogged in mud, and fields mowed then haymaking abandoned, the rows of grass just left to grow in.

'Sa, 'ow many lal' bales 'as ta made?' asked Eric, a farmer from further down the dale.

'Four,' Clive said flatly.

'Thousand?' Eric said, coupled with a look of disbelief

and admiration. 'Four 'undred?' he asked, his reverence duly fading, when Clive remained silent.

'Nah. Four . . . as in four,' I declared dryly.

While the building work was going on at the farm over the summer, at The Firs I was instigating the gradual movement of surplus items from Ravenseat. Given that we needed to furnish a large, six-bedroom house more or less from scratch, it was fortunate that I saw myself as a bit of a collector of objects of historical interest. Over the years, I had amassed all manner of weird and wonderful items. The loft of the old chapel that we used as a woodshed was filled with the overspill from the house, including ornately carved chests, a barrel-backed settle, a clockwork dutch oven, peat bellows and a butterchurn. Finally, it could all go back into the type of house where it rightly belonged rather than sitting gathering dust in the loft. Unfortunately, I didn't have any of the items that my potential holidaymakers would definitely require – sofas, tables, chairs and the like. Those things all needed buying and this was to prove a nightmare for someone who lacks the focus or willpower to agonize over everyday mundane objects.

I had planned to concentrate on a room at a time, im-agining what our guests would need (or, in reality, what I liked and deemed that they should have). Of course, the plans went awry very soon into the job when it became apparent that shopping at an auction house or on online auction sites meant that you had to seize the moment and bid on what was available at that particular point. Many, many items were bought long before the relevant room was ready and had to be shunted around whilst kept under wraps, protected with dust sheets and blankets. The auction house at Leyburn held weekly, monthly and speciality sales but it was the 'general and household contents' sales that drew my attention. I would go on the viewing day, when all the items for sale were on display. This way I could take the children and spend a very pleasant hour rummaging through boxes of oddments and peering into cabinets without feeling under too much pressure to keep the children in check. Although obviously they couldn't run riot, they could have a good time as there always seemed to be something that would catch their eye. It appeared that there was a grading system in place. There was a room teeming with experts, dressed like dandies and scrutinizing things through loupes; this room was to be avoided at all costs. Then there was the mediocre room where we could proceed with caution, and finally there was the random-stuff room where there'd be 'less valuable' objects. In here we were safe, and the porters would point us in the direction of a rocking horse that would keep the children entertained.

I often had no real idea what I was looking at or for, though often the auction guide price written in the catalogue determined whether something was of interest to me or not. In other words, it had to be cheap. If I found something that looked suitable for my project then I would leave a bid on it and hope for the best; this method also prevented me from getting carried away and getting into any bidding wars.

We became familiar with all of the staff at the salesroom, the porters who came ready equipped with measuring tapes and pencils and who would happily corroborate the provenance of any item.

'Is this bookcase solid oak, an' where's it hail frae?' I'd ask, glancing between it and the catalogue.

'Yeah, come outta a gurt fancy spot oo'ert North East an' it weighs a bloody ton.'

It was good to have these lads onside as they would also sometimes let me have a sneak preview of what they had in storage for the next sale.

Rodney was the chairman of the salesrooms and could always be seen buzzing around the building. He was an old-fashioned charmer with tailored suits and the gift of the gab. Quick witted, he missed nothing, and all the time he conversed with people his sharp eyes would be darting here and there on the lookout for potential buyers. He had a needle-sharp memory and could easily recall a conversation had weeks beforehand, remembering all the intimate details. He had, in essence, all the attributes of a great auctioneer.

On one particular morning, I'd been directed to leave the 'shop floor' and go upstairs into the showroom that was reserved for the finer sale items. One of the ladies who manned the front-desk reception had glided over to me after seeing me and the children walking through the foyer.

'You *must* go upstairs and see our latest collection,' she purred. 'In fact, I implore you to take the little ones up the stairway . . . they'll just adore what is on display.'

I was always bowled over by these women, typically whippet thin, legs longer than those on a racehorse and always impeccably turned out, often in bouclé skirts and jackets. Young or old, they all looked achingly smart with sleek, shiny hair styled into low buns, perfectly applied barely-there make-up, dainty fingers and swanlike necks usually decorated with pearls. I wished that I could ooze that sophistication.

Then she beckoned me to come closer and, putting her hand partially over her mouth, whispered in my ear.

'Lot fifty-eight . . . just look at its testicles.'

I didn't need any more persuading to go and investigate further.

There was going to be a taxidermy sale and having previously purchased the Blue Gnu they knew I might find something I liked. The room was a macabre mix of trophy hunters' large-game mounts, antlers and bleached bones. On tables stood domed glass cases filled with exotic birds of paradise, and draped over chairs were skins of zebras, antelopes and deer. The children stared open-mouthed at the grotesque figure of a stuffed monkey wearing a fez and smoking a hookah pipe. I methodically worked my way along the lots, but I knew long before I reached Lot 58 what it was going to be, for laid out on a table of its very own was the biggest dog I had ever seen. It was a truly monstrous sight, a Great Dane stretched out as though lying in front of a fire. His long legs were extended, his head was curled around, his chin rested on the floor, and his eyes were closed. I say 'he' because there was no missing his sex – he was sporting a huge, overstuffed pair of testicles that protruded from between his legs. In fact, they rather drew the eye from whatever angle you looked at him.

Accompanying this most morbid of lots was a DVD: Jeremy Brett as Sherlock Holmes in *The Hound of The Baskervilles*. This poor stuffed creature was cast in death as the hellhound.

Needless to say, I didn't buy anything; it certainly was a

sale for those with an acquired taste. We made our way back downstairs and into the foyer where I saw Rodney talking to a group of tourists. He swung around, throwing his arms open wide.

'Amanda,' he called theatrically. 'Darling, what brings you here today?'

Before I had a chance to reply, he strode over and dropped to his knees. Annas looked on open-mouthed whilst Rodney thrust his hand into his pocket and pulled out a small antique jewellery box. He flipped the lid open with his thumb to reveal a stupendously large brilliant-cut diamond ring.

'Amanda, would you do me the honour?' he asked in a booming voice that echoed through the building, stopping people in their tracks and making them turn to see who the recipient of such a romantic gesture was. I coloured furiously as he held the sparkling ring in his outstretched hand.

'Rodney, I've got a bit of baggage, yer know,' I said. 'Not to mention a husband.'

Then I grabbed the ring.

'Whose is this, anyway?' I said, studying the sizeable rock. 'Elizabeth flamin' Taylor's?'

'It's in for valuation,' he said.

I decided that it was time that I gave him a taste of his own medicine. I shoved the ring firmly onto my ring finger.

'Oh dear,' I muttered, pretending to pull at it, 'it seems to be stuck, Rodney. I can't get it off!'

The look of panic on his face was priceless, as was the look of relief when I twisted it back off.

I was successful in buying the majority of items that I left bids on, though, probably because there wasn't a lot of competition for my chosen pieces. I liked things that were very old, well built and basic, that I could run my hand over and wonder about where it had been and what it had seen throughout its lifetime. Like I did with Clive, I suppose.

I bought an ebonized court cupboard, a brass-handled dresser and an oak coffer, initials hewn crudely into its frontage. They'd been loved and cherished by families for generations then somehow ended up catalogued and lined up around the walls of the saleroom waiting to be auctioned off to the highest bidder. When these items were made, the country cottages in which they stood were furnished meagrely and these cupboards would have been used for the storage of valuables. Indeed, inside the coffer was a built-in candle box, somewhere to keep these most precious and expensive of items.

The older folk in Swaledale remembered a family moving into Ravenseat in the 1920s who brought all of their worldly possessions in just a single journey. One trip in a horse and cart from Orton in Cumbria some twenty-odd miles away with all their family, their belongings and everything they held dear. Then, a few years later, they loaded it all back up again, only this time heading to their new farm towards Barnard Castle. It is true that family heirlooms would have been treasured, but it seems that certainly amongst the poorer folks there was little sentimentality for such things. I talked to an elderly friend who reminisced about clearing out an attic after a family bereavement. Box after box of manuscripts and papers, maps and deeds were all deemed to be worthless and burnt on the fire. I suppose that there is a fine line between today's junk and the next generation's treasure.

The half-tester beds were an accidental purchase. Reuben

and Miles had renovated a pair of iron bedsteads which looked terrific but, with another four bedrooms standing empty, I still needed more beds. Buying them second-hand didn't seem right, so for a change I would be buying brand new. That was until the special 'Country House' sale catalogue arrived, and I leafed through the pages. There, in the section titled 'contents of a gentleman's residence', were two half-testers. The only reason that these two lots caught my attention was that I had no idea what a half-tester was. A quick look online told me they were similar to a four-poster bed but the drapery hung only from a canopy at the head end. The ones pictured online in the antique shops looked fantastic, with swags and drapes and embroidered sunburst panels, and with a cost of over three thousand pounds for one in pristine condition. The catalogue was asking for bids in the region of five to seven hundred pounds, clearly due to the great difference in condition. While the beds in the catalogue were a deep rich mahogany, they were also in pieces, dismantled and disorderedly, with nylon and polyester drapes in a salmon-pink floral pattern, very reminiscent of the 1980s and definitely a fire hazard! The porters assured me that they were complete, for in polythene bags were all of the nuts and bolts required to reassemble the beds.

Clive was not as enamoured with the beds and questioned whether they were in keeping with the decor and feel of The Firs. I reassured him that four-posters and half-testers had actually been more common than folks assumed, and that not only were they to be found in mansions and castles but also in old farmhouses. Standing high off the floor, and with their curtains, they kept the draughts at bay and also afforded the occupants some privacy in rooms that were often shared with other family members. Despite Clive's protestations, I booked a phone line for the sale of the beds; I was dead set on getting them but unsure of how much money I was going to have to part with to buy them.

One of the porters would ring me on the mobile phone a few minutes before my chosen lots came under the hammer, and would bid on my behalf if I gave him the go-ahead. I would be on a train from Darlington to London King's Cross when he called at one thirty, nicely settled in my seat by then and able to talk.

Unfortunately, the train was running ten minutes late. Just as the tannoy announced that passengers for the one o'clock service to London King's Cross should be getting ready to board as the train was arriving at platform 1, my phone rang.

The train slowed as it pulled into the station, and other commuters strode forward. I took the call, wedging my phone between my shoulder and my ear whilst pulling my suitcase towards the jostling queue of people. Children cried, the train's air brakes hissed, the voice on the tannoy reminded people that the train was about to leave and, very faintly, I could hear the patter of Rodney the auctioneer. A voice stated that the next lot to be sold was lot 373, a Victorian half-tester. Another voice asked whether I needed a hand onto the train with the unwieldy suitcase. It was brain overload. I can multitask with the best of them but at this point I was really struggling, as was the poor man on the other end of the phone line. I don't know quite how much he could hear above all the racket or whether he could make sense of the two different conversations I was having. Finally, I got through the door and set off down the central aisle trying to find my seat. Nobody coming in the other direction stood a chance with me like a ship in full sail as I marched along dragging my suitcase, and holding my mobile – now on speakerphone – in front of my mouth like a Ryvita I was just about to take a bite out of.

'I'm on the train!' I shouted as we began to pull out of the station. I steadied myself by turning and leaning against the side of an aisle seat. I'll just sit here for a minute, I thought, get these beds bought and then find my place.

'Three . . .' said the voice on the phone.

'Yer what?' I said, frowning. 'We're goin' through a tunnel.' I could hear the phone line crackling.

'You're breakin' up,' I shouted.

'You're in the wrong carriage,' said a lady sitting opposite me.

'What?' I said.

'What?' the stuttered voice replied on the phone.

'This is the quiet carriage,' said the lady, who I noted had very pursed lips and looked a little annoyed – and rightly so.

'Oh OK, I'm sorry, I'll move,' I said.

'What?' said the voice on the phone. 'Are you in at three?'

'Just buy 'em,' I snapped. 'Just buy 'em both an' then ring me back.'

I switched the phone off and shuffled back down the aisle hoping to hell that I hadn't just bought two very expensive beds. In the end, I paid £350 for one and £400 for the other, so that was a good result but more by luck than by management. Clive wasn't as pleased as I was with my bargain buy.

''Ow much did ta 'ave t' part wi' then?' he asked when I rang him later to tell him that I had arrived safely in London.

'Nay, not as much as I thought,' I said. Trying to lift his mood, I added, 'an' Reuben will love puttin' 'em back together, there's no instruction manual tha' knows.'

I was itching to go and collect them and arranged with the salesroom that I would come and pick them up in a couple of days. I figured that the cattle trailer would be big enough to get all of the components in, and if I took some baler twine, the farmer's friend, I could easily secure them so that they didn't slide around during transit.

Clive helpfully put the cattle trailer on the back of the pickup and parked up outside the front of the house.

'There yer go mi darlin', trailer's on,' he said cheerily. After all, the sun was shining, the grass was growing and the yows and lambs were content.

'I'll see yer later. I'll help thi unload 'em,' he said, waving as he strode off up the yard. I loaded Annas and Clemmie into the pickup and off we went, windows wound down, singing away and chattering about our plans for the summer. In little time at all, I pulled up at the salesroom building and followed the signs for deliveries and pick-ups, joining a queue of courier vans and estate cars. We sat waiting for ages. I turned the engine off and we played I-spy.

'I spy with my little eye, something beginning with . . . "P".'

Annas would come up with random suggestions usually not even beginning with 'P' but she was still just three years old.

'Painting, Annas,' I said, pointing to the porters manhandling a bubble-wrapped oblong object towards a waiting transit van, 'probably Picasso.'

I patiently waited, watching crates being wheeled onto the loading docks and official papers being signed. Finally, after a few stops and starts as we edged ever closer to the front of the queue, it was my turn. I swung the pickup and trailer around, opening the driver's side door to get a better view of what I was doing and backed the trailer up to the loading bay.

I got out and went to drop down the trailer's back door. The porters were already bringing the beds in pieces towards the docks.

'In t'trailer, yeah?' said one of the lads.

'Yes, please,' I chirped as I opened the fold-out gates inside, but then I was confronted with a picture of horror, for the inside of the trailer was awash with sheep poo. Not just a few dried-out currants neither; the inside of the trailer was coated with lashings of the stuff, still wet.

'In there?' said the porter with an incredulous look on his face.

I groaned. Clive had obviously had sheep in the trailer and I hadn't thought to check whether it had been washed out before I set off. I had assumed that it was clean.

The only real remedy was a hose pipe, shovel and brush but not having access to any of these things left me with no option other than to put the top level of decks down inside. These were what we used if we needed to carry sheep both upstairs and downstairs. Fortunately, the upper deck was in a better state than the lower, though it still meant me walking

161

through all the muck to lift all the catches and drop the fold-down chequer-plate floor. The porters, bemused by the state of the trailer, found sheets of cardboard to protect the furniture a little and eventually loaded it all onto the top deck.

I chastised Clive when I got back home but, quite rightly, he pointed out that it was my job as the haulier to see that the trailer was fit for purpose, helpfully adding on an old favourite quip:

'S**t 'appens, Mand.'

I vowed never to make that mistake again.

Reuben desperately wanted to reassemble the beds but, knowing that they required a little alteration in order that they stood level and square in distinctly unlevel, off-kilter bedrooms, I asked our friend Alec to come and help. I knew that Alec, with his eye for precision and detail, a quality that had annoyed Clive on innumerable occasions, would not tolerate any imperfection in the reconstruction and so would keep Reuben on track. I couldn't imagine anything worse than the whole bed and canopy above collapsing on an unfortunate couple in the throes of passion.

This was the last time that I bought anything at the sales-rooms, having come to the conclusion that I was making things difficult for myself. The beds, although uniquely wonderful, took such a great deal of time and effort to make usable. My friend Rachel is an excellent seamstress and volunteered to make me new drapes. Ken, the joiner, fixed new headboards in place and rather than buy new, very expensive made-to-measure mattresses for the non-standard-sized frames, I made use of some very heavy and long feather-down-filled bolsters. The girls adored the beds, the velvet throws, silk tie-backs and damask cushions that were very reminiscent of the Princess and the Pea fairy tale, and they all wanted these beds to be theirs. It was this that made the whole undertaking worthwhile, beds that are heirlooms to be cherished and hopefully stay on in the house for generations, long after we have departed this earth.

It was now Ian, from the antique shop, that I went to with a wishlist of items that I needed for the house. A large farmhouse-style table to seat at least twelve people, nothing too highly polished, just sturdy and serviceable and certainly not too expensive.

'They're not easy to find, farmhouse tables,' Ian told me, 'and big ones, well they're a rarity indeed.'

I also needed a coffee table, a couple of chests of drawers and some kind of side table for the kitchen. Again, nothing too spectacular, I was open to all ideas. I heard nothing from him for weeks, then one sunny afternoon I was driving up the one-way cobbled street in Hawes after selling sheep at the auction when, as usual, I slowed to study the window display in his shop. Ian was standing in the doorway watching the world go by and, seeing me crawling past, gestured that I should drive back around the one-way block again. By the time I was in front of his shop, he'd swung open the back doors of his transit van on the side street and was pointing enthusiastically at a table inside. I did an emergency stop and parked up with my hazard lights flashing.

I walked over to the van for a closer look.

'Do yer want mi to get it out for yer to 'ave a proper look at?'

I told him that there was no need and that if the price

was acceptable then he might just as well leave it in the van and bring it to The Firs.

'Well, I had to go to Keighley for it,' he said, 'and it is a very nice table in original condition.'

'Yeah,' I said.

'Seven hundred pounds and I'll throw in four chairs an' all.'

We shook hands on it and agreed that Ian would deliver the table the next day. New reproduction kitchen tables that I had seen online were more money than that, and I was getting the genuine article. I was very pleased indeed.

I told Clive that his presence at The Firs the following afternoon would be much appreciated as there was a kitchen table being delivered and it was going to be heavy and undoubtedly awkward to get through the hallway and into the kitchen.

'Yer know, I dreamt last night that we're nivver gonna get this table through t'door,' I said, waiting until we were on our way down the road to The Firs before I decided to share this information.

There have been occasions when, in an aspirational mood, I have told the children that their dreams can come true, and unfortunately on this particular occasion I was right on the money. The table was just too big to navigate the turn into the house from the front door. Clive and Ian even tried carrying it right round the back of the house and through the field and attempted to gain access via the sunroom, but that didn't work, either. It was a very hot day, tempers were getting frayed and there were even some mutterings from them both about a refund and taking it back to the shop.

'Well, I reckon we could get it in if we sawed t'legs off,' said Clive, mopping his brow whilst Ian leant against the wall getting his breath back and looking down on the table, now lying sideways, half in and half out of the sunroom.

'I've never sawed the legs off an antique table,' said Ian, slightly indignantly. 'In fact, I've never sawn up any kind of furniture in order to get it into a house.'

'Well, there's a first time for everything,' said Clive. 'It's the only way this thing is gonna get into this 'ouse.'

'No, I'll take it back,' said Ian, clearly mortified at the idea of desecrating something old.

'Naw, we need to be brutal, let's just get on wi' it,' said Clive and, with that, he went out to the garage, returning moments later with a hacksaw. I cannot say that I was entirely happy about the extreme measures being undertaken, and I swear I saw Ian wince as Clive made the first cut. As Clive sawed, cleverly, at a forty-five-degree angle to make the repair job afterwards easier and stronger, he hit a sizeable iron nail.

'Why, that's a capper!' he said as he examined the partially sawn table leg. 'Ach, it's been sawn down afore.'

He was right, all four legs had previously been amputated, probably for exactly the same reason, then duly reattached with an almost invisible mend. I left it in the capable hands of Ken the joiner to do the repair job, and the table now stands in the middle of the kitchen, the scrubbed top pitted and gouged, the boards worn and slightly dipped at the edges. It is a piece of furniture that will probably never move now, or at least not without undergoing further surgery.

The other items on my wishlist were sourced and delivered

without mishap, an elm dough table for the kitchen which, with its coffin-like construction, provided a perfect place for tea towels, tablecloths and linen. The coffee table was supremely heavy but not difficult to move owing to it being on iron wheels. Of course, it didn't start out life as a coffee table, far from it, it was one of two mining bogies that Ian had found . . . well, in a disused mine, I guess. One is a deep-sided tub made entirely of metal that sits outside the house full of logs for the fire. The other is low, wide and made of oak, with riveted protective metal edges, and is the perfect low table, though only after Reuben made chocks to prevent it from rolling away across the flagged floor. The children were fascinated with the carts and would play happily with them, pushing one another around in them and taking each other for rides, but what pity I had for the children and ponies that had been forced to spend their lives in the coal mines at Tan Hill and the lead mines of Lonnin End just a stone's throw away. Hauling these heavily laden carts along the underground passageways and then to the surface, for little recompense, must have been dangerous, back-breaking and soul-destroying work.

People round and about soon got to know that I was on the lookout for furniture and smaller pieces to fill a house. I was astounded by people's generosity but also surprised at how wasteful society could be as a whole. Pictures, curtains, bedspreads and cushions all surplus to requirements due to redecoration and a change of colour scheme came my way. A friend brought box after box filled with blue-and-white Spode tableware: meat dishes and platters, teacups and saucers and large lidded serving dishes. A rustic earthenware pancheron was retrieved by Peggy, one of our neighbours, from an old dairy at Angram. A large dough bowl, its name is derived from the process of kneading the dough in a circular motion, or panching. The same description is often used to describe an animal going round and round in circles before giving birth. Apparently even I have been known to panch! This particular pancheron had been thrown by hand on a pottery wheel then glazed and fired, the ridges made by the potter's fingers preserved forever. Its imperfections are what make it so unique and beautiful to look at. It had stood unused on a stone shelf for too long; though still fit for purpose there are now not so many who would need a dough bowl of such large proportions.

Unfortunately, I was also given items that I truly did not want nor care for. Sometimes it was a case of exaggeration on the part of the bestower of the gift.

'It's a grand owd chair, mebbe could do with a bit of a clean-up but it's a bit o' age.'

The reality was a 1970s monstrosity which, rather than needing a clean-up, needed to go in a skip. I sound like such an ungrateful beneficiary, but you'd understand if you saw the things. In the end I just found it easier to accept the gifts rather than refuse and hurt someone's feelings.

On one such occasion I was given a larder cupboard, nothing of any great age but evidently it had once been quite a nice piece of kitchen furniture with full-length doors, a bottle rack at the bottom and shelves at the top. It had stood in an outhouse for a few years after the adjoining house was cleared out after the owner's death. It was still being used as a larder cupboard, only now it was storing dried dog food, the smell of which had permeated the wood. Reuben worked his magic on it; he sanded it down, stained and waxed it and when it was finished it looked very much in keeping with the farmhouse kitchen. It had looked irredeemable when I first set eyes upon it but I was so pleased with the end result that I decided that Reuben and I should show Bryan, the donor of the cupboard, the transformation.

Bryan was impressed. It was nearly unrecognizable, only the faint smell of dog food giving the game away.

'I don't suppose you want a dresser?' he said. 'Nice thing, made of oak, proper craftsman-made but we've no need for it now, it's too big, we just haven't got room for it.'

I can honestly say that I really didn't want a dresser, but it seemed impolite to turn down the offer and it didn't help that Reuben, ever the magpie and collector that he is, was nodding his head vigorously and already making plans for this next project.

'Strike whilst the iron's hot. Follow me back home now,' said Bryan, 'and I'll help you load it.'

I couldn't think of any viable reason as to why I couldn't, so off we all went. I could see why Bryan didn't have room for the dresser, it was enormous, solid and extremely heavy. Reuben set to work with a screwdriver and removed the plate rack from the back which made it slightly less cumbersome, but still it took some moving. We edged towards the pickup, walking it ever so slowly, using tiny steps, and all the while I was thinking how much I hated the flaming thing. Once we got it to the pickup, it was a question of lying it down and anchoring it, and the now-separate plate rack, in place.

'Don't want it to move,' said Bryan as he tightened some

elastic bungee cords around the tailgate. I did have a very steep hill to negotiate on my way back home.

'No, that wouldn't be good,' I said. 'Perish the thought.'

Reuben knew me only too well and could see through my mock gratitude.

'Yer don't like it, do yer?' he said as we drove back, taking the twists and bends that Swaledale's roads had to offer a bit quicker than usual, secretly hoping for there to be some 'accidental', irreparable damage in transit.

'No, Reuben, I'm afraid I don't,' I said flatly as I flew into another chicane. In the back I heard a dull thump as the dresser slewed against the pickup side.

'Are we gonna unload it tonight?' asked Reuben. I told him that I wasn't, and I really wasn't that bothered if it came out of the pickup in pieces.

Clive came out to see what I'd salvaged this time.

'Oh, now that is lovely,' he said, running his hand over the wood. 'Proper quality, nice thing.'

Reuben said nothing and just looked at me despairingly.

Clive then insisted that we took it straight down to The Firs. When, finally, after an almighty effort, we had moved it to its final resting place in the sitting room, Clive stood back to admire the monstrosity.

'There,' he said. 'Looks well, don't it.'

He must have noted the look of disdain on my face.

'Personally,' I said, 'I hate it, it's horrible.'

'It'll grow on you,' argued Clive.

I left it at that, there was no way that I was going to expend any more energy on it that night. It could stay until I found something better to take its place.

Annoyingly enough, from that moment onwards, every-body who visited The Firs would comment on that dresser and it irritated the hell out of me. 'Come on,' I'd say. 'Follow me and see the wondrous and interesting things that I'm filling the house with.' We would wander from room to room whilst I pointed out what I thought were objects of fascination and beauty and finally, when we'd taken in the panelled porch and chandelier, the half-testers and polished back range, they'd stand and gawp at the dresser and tell me what an outstanding piece it was.

Reuben polished it and replaced a few of the cup hooks on the rack but it made no difference to its overall appear-ance, it just stood dominating the room with its angular, awkward presence. It only just fitted in the gap between the beams. Once upon a time a drop-down cupboard bed had sat in the very same place, a feature from times past when it made no sense to leave the warmth of the fire to go upstairs to bed. Unfortunately, this had been removed and now all that remained were the old nails in the beams and wooden pegs in the flags that had once underpinned it.

I had just got to the stage of considering whether it would be acceptable to take the dresser to the saleroom, when I had a stroke of luck. Our friends Colin and Anne from Weardale had called on us and we had once again taken them down to The Firs to proudly show off our work in progress. They were interested in how we were getting on as they were about to embark on a similar project themselves, converting part of their farmhouse into a bed and breakfast and holiday let. I dutifully showed them around, all the nice places, rooms both finished and unfinished, and then ended the tour in the room with the French stove where the dresser stood. I opened the door and, from the doorway, just vaguely pointed out the colour schemes and explained how, one day, eventually it was to be filled with bookshelves, prints and maps but, in the meantime, it was just being used for storage and a workspace.

'Nothing to see here,' I said to them, not wishing to encourage them to venture in and undoubtedly express their admiration for the ugly dresser. Of course, Anne was having none of it and went in for a closer inspection.

'Col, come an' look at this,' she piped.

I sighed inwardly, but it was fortunate that I didn't share my avid dislike for the dresser aloud because Colin turned out to be an aficionado of the cabinet maker who had made it some thirty years before.

'Aye, it's made by the same company that made our kitchen,' he said. 'Look, dovetail joints and the same lal' carved acorns that were his signature mark.'

'What a find,' said Anne. 'I'd love one just like it.'

'Yer can 'ave this 'un,' I said. 'Believe me, you'd be doin' me a favour.'

'Now hold on, Anne,' said Colin, getting flustered. 'We haven't got the room, there'd be nowhere to put it . . . unless we get rid of yer Mam's pipe organ.'

'I can't get rid o' that, it was Mam's pride and joy,' she replied indignantly.

I smiled. A solution had just come to mind that would keep both parties happy.

'A pipe organ. Yer know what, I would rather like to own a pipe organ.'

'Yer would?' said Colin.

'Yer would?' echoed Clive, who had now appeared on the scene.

'It's a big un,' said Colin.

'Mand'd accept nothing less than a big . . .' Clive smirked, but I didn't let him finish.

Anne was very keen for this swap to happen and came up with a plan. Colin would take a picture of the organ on his phone and would show me when we were both at Muker Show the following week. Mobile phones with cameras have now gone mainstream to the point that even the most hardened of Dalesfolk will have one lurking in their pocket or in the glovebox of their Land Rover. It is irrelevant that, for the most part, there is no coverage so little chance of actually being contactable, but what they *are* useful for is the photographing of sheep. At the Muker Show they were perfect for taking pictures of the day's show winners or, even better, the sheep that you had at home but that you didn't bring to the show that day but would almost certainly have wiped the floor with the other competitors. To the untrained eye it all looked mighty suspicious to see small gatherings of archetypical Dales farmers huddled around a mobile phone. Whether it's a relief when you discover that the screen they are studying does indeed show a lithe, long-legged female – but rather the perfect embodiment of a Swaledale sheep – remains doubtful.

'Now what yer looking at?' asked Stephen Calvert, the builder, as I squinted at the picture on Colin's tiny phone.

'A picture of his organ,' I said, keeping my face perfectly straight.

'It actually is an' all,' Clive, who was leaning against the Land Rover, chipped in.

The conversation soon got turned around to quoits, Stephen's hobby and lifelong passion. He had scaled the heights of stardom in the world of international quoit playing, playing all the big venues on the circuit: Arkengarthdale, Whitby and Goathland. Muker Show was one of the most prestigious locations to play hoopla – whoops, sorry, quoits – on the hallowed turf.

'Why aren't yer playing?' enquired Clive, slapping Stephen on the back. It did seem unnervingly quiet without the familiar chink of the metal hoops hitting each other after being thrown towards the wooden peg in the clay pit.

'Nah, no quoits competition this year,' said Stephen, shaking his head sadly. 'It just couldn't go on any more.'

'Why, what's happened?' I asked.

'It were just becomin' *too* popular, folks from all ower turning up to play.'

'Nay to hell, yer can't be 'avin that,' quipped Clive, 'folks turning up at Muker Show and wanting to enjoy themselves and play quoits? No wonder yer put a stop to that.'

The pump organ was beautiful and ugly in the same breath, made of mahogany, over-embellished with fretwork and an arrangement of large, painted, decorative pipes. Textured and weirdly tactile, they resembled giraffe skin, though the colour was not dissimilar to verdigris. Above the keys was a row of stops with glorious musical terms written upon them. A pair of velvet-clad pedals needed to be rhythmically pumped in order that the internal bellows filled with air and the resulting sound was as powerful and stirring as it was captivating.

'No wonder yer mam loved playing this,' I said to Anne after we finally had it in the place of the dresser, which was now loaded up and awaiting transportation to its new home. 'Are you sure yer wanna part with it?'

'This is the next chapter in its history,' she said as she reminisced about her mother Gladys's love of music. She had been an accomplished pianist and teacher, having trained at the London College of Music, and for many years had held organ recitals in her sitting room and organized sing-songs at the local pub.

My pathetic attempts at making sweet music were a miserable failure but, on the upside, the resistance I met with when pumping the foot pedals would at least give me toned thighs.

'There's many a good tune can be played on an old fiddle,' commented Clive, 'but that aint one of 'em.' He winced at the disjointed few bars of 'Beautiful Dale (Home of the Swale)' I hammered out.

'Mebbe I should be practising this, I might be needing to play it soon,' I replied. I picked out the sombre tones of the funeral hymn 'The Day Thou Gavest'. 'How d'ya like this?'

Clive shook his head. 'If you don't want me shuffling off this mortal coil just yet, mi dear, yer need to be keeping me in the manner to which I'm accustomed.'

'You'll live forever, thee,' I said glibly. 'Creaking gate an' all that.'

6

Where's Eartha?

'BARN SALE' the advertisement said, the rural equivalent of a garage sale. The listing in the local free advertiser read like a dream: cheese press, hay spades and other farming bygones, it said, then a telephone number. I rang straight away.

'When's the sale?' I enquired. 'It doesn't say in t'paper.'

'Why, what's ta wantin'?' came the curt reply.

I explained that I didn't know until I'd seen what was for sale.

There was a long silence.

'We'eell thoo'd better come now,' came the reply, eventually.

Swaledale has many little green lanes that veer off into the unknown, tracks hemmed in by drystone walls that inhibit the passage of anything wider than a four-wheel drive. These lanes tend to follow the contours of the land, winding their way to secluded farmsteads, and it was one of these lanes that I was now carefully negotiating in the Land Rover, cash in pocket and excited children in the back.

We had already had a stroke of luck on the journey from Ravenseat, so spirits were high. The road down Swaledale was winding, with hairpin bends, humpback bridges and numerous undulations that could test the skills of any driver, particularly any that were loaded to the gunwales with merchandise. White van men, given an impossible timetable in which to deliver parcels to houses in the area, would undoubtedly test the protective packaging of those items marked as 'fragile' or 'handle with care'. Only recently, I was sent a gift of a very pretty religious ornament of the Good Shepherd complete with a pensive look as he scanned the horizon, presumably looking for his missing lamb. Unfortunately, it wasn't only a lamb he was missing when he arrived: a limb, too, had become detached, presumably lost in transit.

Today, though, the bumpy road was to be a blessing, as it was our turn to be the finder of good things. Just as we rounded the corner out of Gunnerside, I spied something lying ahead of us in the middle of the road.

'What's that?' shouted Sidney, who was travelling in the middle seat on the middle row, looking over my shoulder.

'That, my little ginger freckled friend,' I said, 'is a loaf of bread.'

I checked that there was nothing behind and then did an emergency stop. Reuben leapt out of the back door and scuttled off to retrieve it.

'It'll be all right for thi chickens, Miley,' I said as Reuben got back into the Land Rover.

Reuben reported that the bread, medium-sliced Warburtons, was in date and the packaging unscathed.

'OK, it's not for t'chickens, it's for us,' I informed Miles.

'Do yer think we should hand it in?' said Violet innocently. 'To the police.'

'Use yer loaf,' I said. 'Finders keepers, in this case.'

It was like a modern-day version of Hansel and Gretel following the trail of breadcrumbs, only, in our case, it was Warburtons toastie, medium-sliced or – the absolute pinnacle – a seeded granary. Every few hundred yards, there'd be another loaf. This went on for a couple of miles until the trail petered out. Whether it was a bread van or just a supermarket delivery vehicle, I don't know, but it kept us in sandwiches for a few days.

And so, the children chattered away and played pass the parcel with the loaves of bread as we jolted and jarred our way along the narrow track.

'Are we nearly there yet?' piped up a little voice but, quite truthfully, I didn't know, as I was in unfamiliar territory and following the vaguest of directions. Finally, we rounded a corner and the lane opened out into a farmyard. The weather was dry so, devoid of mud, the once-cobbled yard looked charmingly quaint and in no time at all the children had disembarked and gone. Only the little ones remained, baby Nancy now wide awake and squinting in the sunlight and Clemmie, who had a perplexed look on her face, was straining against her seatbelt in an attempt to free herself. The children had no qualms about setting forth to explore the gardens, fields and pastures; in their eyes nowhere was off limits. Being raised at Ravenseat, in open country where you can walk for hours without ever seeing another soul or passing habitation, the concept of not being allowed to roam wherever they liked never occurred to them. I wasn't going to be the one to tell them that it didn't always work like that.

There was certainly plenty to see. Cats and chickens featured predominantly, as, too, did various tractors, vintage and otherwise but all in a serious state of disrepair. The overall picture was one of a place that had seen its heyday and was now in serious decline, but still it was a pleasant enough scene. The fact it was a fine sunny day is probably what made it seem homely and welcoming, for on closer inspection it appeared that no scrap man had set foot in the farmyard in at least the last fifty years.

An elderly man emerged from a building and shuffled towards me, his head bent forward, a greasy cap upon his head. His heavy black wellies were folded down, so the frayed hem of his ill-fitting brown moleskin trousers was visible. Braces appeared to be holding everything up. His shirt was dirty, missing at least the top three buttons, and his sleeves were rolled back to the elbows, exposing crêpe-like weathered skin. Here and there, a purple bruise was visible. Old age and infirmity comes to us all and, knowing nothing of him or his circumstances, all I could perceive was that he seemed to belong to another era.

'It's a lang time sen bairns was laikin' in t'yard,' he said, when finally he reached out his hand to shake mine. 'The name's Ambler by the way.'

The children were now back, squabbling and creating a ruckus. The cats that dozed here and there eyed them suspiciously with the withering look of disdain that only a feline is capable of. Chickens that were quietly scratching in the yard now sprinted away into the distance, their peace rudely interrupted.

'This place is ace,' shouted Reuben, who was in his element, poking about around the skeleton of what once was some kind of jeep. The chassis was exposed, and its body had rotted and rusted into oblivion. The remains of the rubber tyres had perished and it now sat forlornly, crumbling and exposed to the elements. Evidently no one had realized that it was on its final journey and the key had been left in situ, ready for the next outing that never came, until it had corroded and become one with the ignition.

'Oh my gawd,' shouted Edith. 'There's an ice-cream van too!'

Cue a stampede as the children ran to a nettle bed where the abandoned van sat. I felt a bit embarrassed at the children's enthusiasm for what was essentially a junk yard. Much of the stuff was way beyond repair, but I was still hopeful that somewhere, in an outbuilding perhaps, there'd be something of interest.

'Nut everything's fur sale,' Ambler said. I was thankful for small mercies; Reuben would have happily taken everything he could lay his hands on.

'Com wi' me,' he said, ushering me towards the farm buildings.

Raven was sitting on an old stone mounting-block, tickling a tabby cat which had stretched itself out and appeared to be enjoying the attention.

'Rav, watch lal' uns will yer,' I shouted as I followed Ambler.

We stepped over pushbikes and chain harrows, skirted around milk churns and dodged a back actor (an excavator that is mounted on the back of a tractor) precariously balanced on blocks. He stopped at an ancient building with an arched entrance. Once upon a time, it had had doors, but not any more; hurdles had now been tied across the front to stop animals getting either in or out – which, I wasn't quite sure.

'There's some stuff in t'granary,' he said.

There were some useful items laid about – muck forks, a spade and a shovel – all solid and well made, as things used to be. Anyway, I owed Stephen Calvert the builder a shovel, as I'd managed to run his over and break it when we were working down at The Firs.

'What's 'appened 'ere,' he'd exclaimed, although the wheel marks and the close proximity of the Land Rover meant that it wasn't really that much of a mystery.

I apologized. 'It wasn't your favourite shovel, was it?' I said.

'Aye, it bloody well was,' he replied forcefully.

I grimaced, not entirely sure whether to believe him.

'Forty years I've 'ad that shovel, man an' boy,' he said. 'I've replaced t'shaft an' even bought it a new end, but still.' Then he laughed.

Now, at the barn sale, I found a gavelock, a long iron bar for driving holes into solid ground and a tiny moudie spade perfect for breaking into the tunnels when putting down mole traps. I paid him, and he carefully counted out the notes, then folded them and put them in a dinted metal tin in his pocket.

'I was really looking for some household stuff too,' I said. I was itching to see what else he might have.

'Aye, that's what t'last fella was lookin' for an' all,' he replied.

Of course, I wasn't the first person to go and have a look at what was on offer, news travels fast up the dale and I was still usually the last to know.

''E got t'cheese press.'

I tried to hide my disappointment.

'You had a cheese press?' I said flatly, like it mattered now.

'H'aye,' he said, with a sharp intake of breath. His eyes took on a wistful look as he described how his mother had made the best cheese. How the cheeses were wrapped in muslin and stacked on traves – the wooden shelves in the dairy – and what flavour, like nothing nowadays. I smiled as he recalled milking a particularly quiet cow by hand in the field and carrying the fresh milk back to the farmstead in a backcan. Then he snapped out of his dream.

'I knaw where t'backcan is,' he said jubilantly. 'I remembered where it were.'

We set off yet again, round the back of the buildings, past a partially grassed-over midden, through a passageway with a tin roof, between two buildings and eventually reached the farmhouse.

'Carful,' muttered Ambler, 'she's sittin'.'

'She' was a goose, sitting nonchalantly on the doorstep just under the eaves of a tiny gabled porch. I wouldn't say that I am fearful of geese, but I am well aware of their reputation for being fierce creatures and an excellent alternative to a guard dog. Ambler went inside, and I hovered outside, one eye on the goose and the other drawn by the very dingy and dark entrance hall just beyond. Then I followed him in, eyes focused straight ahead. I strode past the goose, willing there not to be flapping wings and a honk preceding a shower of pecks.

Once I'd safely negotiated the feathered fiend I could relax and take in my surroundings. It took a moment for my eyes to adjust to the dimly lit hallway. I was in a narrow corridor, with an arched window straight ahead of me that threw a little light onto the situation. An umbrella stand, crammed full of hazel sticks, stood behind the door, whilst a sizeable banjo barometer hung from the wall above a dresser that was stacked high with envelopes, letters and other correspondence. Creeping black mould had tracked its way upwards from the dampest recesses and now the yellowing wallpaper was peeling at the edges of the skirting board, and flaps hung loosely just below ceiling height.

'Nah need to tek thi boots off,' came a shout from around the corner.

There'd be no danger of that – nobody had *ever* taken their boots off to come in here, that was plain to see. A well-worn path of farmyard muck trailed straight up the centre of the hallway; nobody, it seemed, had ever wiped their feet.

I walked into the kitchen, which was dominated by a hugely impressive stone mantelpiece. It had clearly once sat above a substantial fireplace, though now, in the space that it had once occupied, was a freestanding electric cooker. A hideous pale-blue larder cupboard with vertical-striped frosted-glass frontage appeared to be where foodstuffs and condiments were stored – everything from ketchup to mustard to jars of jam, pickles and cat food were all here within reach of anyone who was to be dining at the kitchen table. The kitchen itself, to give it its due, did contain all the essentials that any kitchen should have: a large butlers sink piled high with crockery, a fridge stood alone by the door and a truly magnificent display of eggshells stacked with tremendous precision and great attention to detail like a croquembouche. There was plenty of seating for any guests that should drop by, but one would first have to clear away the excessive piles of reading material which were stacked and strewn on every surface. Any other available space was filled with biscuit tins or cats.

Ambler was now boiling a kettle that sat on a side table amongst books, newspapers, magazines and newsletters.

'Sit down,' he said, not looking towards me.

There appeared to be, hidden under years' worth of newspapers, a small sofa at one end of the room and, just as I contemplated whether I should begin clearing them aside or throw caution to the wind and sit right down on top of it all, a maelstrom of papers went flying into the air. An elderly man was now sitting bolt upright.

'Aye,' said Ambler, now turning around, 'mi brother, Walter.'

Walter brushed himself down, then, after a moment or two of rummaging around in his pockets searching for his glasses, held out his hand to greet me.

''Ow do,' he said. I sat down on the sofa amongst the papers, thankful that I hadn't just sat right on Walter.

A cat meandered past, arching its back as it stopped momentarily to rub itself on Walter's legs.

I sat, drank tea and enjoyed being in what to many might have seemed like a pretty dire place but I felt privileged to be in the company of a dying breed, two elderly bachelor brothers who remained almost untouched by modern-day life. No television, I learnt, but instead a wireless that kept them abreast of the 'goings on' in the world and, of course, reading material. Neighbours put aside their newspapers for the brothers, and subsequently they hoarded them. As for the books, from what I could see they appeared to be mainly hardbacked and heavy, with faded gilt titles, torn spines and worn covers covered with dust and stains.

'Yer like yer books,' I said as I stood and looked around for a vacant surface, somewhere to put my empty mug. I saw a cat lapping from the milk jug beside the kettle and decided that it probably didn't matter where I put the mug down.

'Yis, a booook is 'ard to beat when it com's to propping a teable leg,' said Ambler dryly.

'Or skededdlin' yon dog,' added Walter.

'I better be making tracks,' I said. 'Lordy only knows where those children have gone, they could be up to naw good.'

Ambler, possibly spurred into action at the thought of me leaving without parting with any more cash, shuffled away back into the hallway and returned clutching a backcan.

I bought it, a beautifully constructed tin can, its wide, rounded body tapering to a funnel-like top, complete with screw-on lid and leather harness to enable it to be worn on the back of the dairyman.

'Aw remember ga'an wi' mi mother up t'field, to Paradise yonder,' Walter said wistfully.

'We used to drink warm milk out o' t'backcan lid whilst she milked the coos, dosta remember?' said Ambler.

I left them talking together in their thick Swa'dale dialect, momentarily lost in their cherished memories.

Back out the house, I sidestepped the goose and made my way through the warren of buildings, tin sheds and outhouses until I was back into the yard. Raven was listening to the radio in the Land Rover, and Clemmie and Nancy were asleep in the back, faces reddened, perfect little rosebud lips pursed, a picture of contentment.

'Where are t'others?' I asked.

Raven shrugged. 'Dunno.'

Ambler and Walter now both appeared, Ambler pushing a barrow in which my newly purchased farmyard tools were laid, whilst his brother, who now I could see was profoundly lame, hotched along leaning heavily on a stick. I whistled loudly to summon the rest of my brood, and they appeared, blackened and grinning, from behind a building, Reuben carrying Annas on his shoulders.

'There's an anvil over there by t'wall,' he shouted, letting go of one of Annas's legs to point out where he was meaning.

'Used to 'ave Gallowa's,' said Ambler.

'Aye, Dales ponies,' added Walter.

''Ad it for when t'smithy com. You interested in t'anvil?'

Of course I was, and once I had seen the glorious depth of patina that had developed over time, run my hands over its hammer-marked form and seen the tiny dirt-encrusted indentations, I knew that it would be a perfect gift for a husband whose only methodology when it came to repairing anything was to hit it repeatedly and hard.

By the time I walked back to the Land Rover, I could hear wailing. Raven later told me that Walter had peered through the back window and frightened the living daylights out of Clemmie, who'd sleepily opened her eyelids mid-dream and been confronted with an unfamiliar face at close quarters.

Loading the very heavy anvil, which was set on a sleeper, was problematic. Finally, after a lot of grunting and a collective effort by Reuben, the elderly brothers and myself, we managed to get the anvil into the back of the Land Rover. Unloading it at Ravenseat was easier as I just backed the Land Rover up to the sloping fodder gang where we fed the cows and, when the back door was level with the floor, we slowly slid it out. At the children's insistence, the anvil was wrapped in decorated paper and bound with sellotape – although the shape made it obvious what lay beneath the layers.

Clive liked his gift and swore that he'd keep it forever, though frankly, owing to its sheer weight, he'd have to be pretty determined to actually get rid of it.

The backcan went down to The Firs, where I hung it by its strappings from a nail in the dairy. A wipe down with WD40 had brightened it up and loosened the screw lid on top. It was just the kind of object that I liked, functional but beautifully crafted and made to last. The end-over-end butterchurn that stood alongside it was an unwieldy object but, for the same reasons as the backcan, it had been brought out of the woodshed loft as an item of interest and decoration that was quite at home in the farmhouse dairy.

I was still in need of one object, the hardest, most elusive item on my list, the cheese press. I saw myself maybe, one day, actually making cheeses. But perhaps this is a similar dream to the one where I make my own yarn from wool: the spinning wheel sits hidden behind the settle at Ravenseat, awaiting the day that I have more time to spare.

It took a while, but eventually I found a cheese press when a small farm was put up for sale, the details going online. The owner had died, and the executors of his will were in the process of clearing out his possessions when they found a cheese press in an outbuilding. I rang the contact number, paid immediately without viewing the item, and organized a courier. I was determined not to miss the opportunity to buy a rare piece of social history. It was an awkward, unwieldy object and only when the transit van pulled up outside The Firs did I really discover the true weight and scale of it. Fortunately, Ken the joiner was in residence, replacing door knobs and putting the finishing touches to the kitchen cupboards, so he was able to give us a hand to unload it and sidle it towards the kitchen. The courier was bemused and stated that it was probably the oddest item he had moved in a while.

'Looks like an instrument of torture, if you ask me,' he'd said as I turned the creaking iron wheels on the threaded-iron bars.

'Looks like it belongs in a museum to me,' said Ken.

It was incredibly well made, the base beautifully and ergonomically shaped from a thick solid slab of oak. Being a double cheese press it had two iron upright poles on which round oak discs sat beneath cast wheels. These were turned and eventually tightened down upon the wooden moulds in which the cheeses were formed. This would squeeze the liquid from the semi-solid mass. The wooden cheese moulds had been carefully constructed to be an exact fit for the oak discs, and imperial weights suspended on chains provided the necessary force to maintain the correct pressure. All in all, it was a feat of engineering, simple, but as precise and effective even after a hundred and fifty or so years of use. It was as good as new and, with a bit of oil to rid it of the squeak and some furniture wax to revive the wooden base, it became both an item of furniture and a piece of farming history.

Unusually, Clive was impressed when he saw my latest purchase, though less so when he realized that it was down to him and Ken to lift it through the narrow doorway and into the dairy.

'Do you think that it's safe?' he asked. 'Yer know, folks wi' kids an' stuff.'

Literally within moments of me dismissing his worries came the scream that said Violet had attempted to crush Miles's hand beneath one of the discs.

There was, of course, always going to be the opportunity for mischief with the cheese press but, as far as I could see, any incident would be a very slow and deliberate attempt to injure rather than an instantaneous accident.

'I think we could make cider with it,' declared Reuben.

'If only we had an orchard,' I reminded him.

'We've got rhubarb, goosegogs and blackcurrants,' he said.

I imagined the resulting potent liquor and decided that it really was not time to start our own brewery.

We were full of excitement and trepidation in equal measure when, finally, the renovation of The Firs was complete, and our first guests were due to check in. One of the big selling points with the house was that it was pet-friendly. As animal lovers ourselves, we wanted our visitors to be able to bring their pets on holiday too, and it just so happened that our first guests had brought with them an enormous, strapping great dog, the proportions of which had to be seen to be believed. When the party, who had driven all the way from Germany, arrived at Ravenseat to meet us and pick up the keys the children were taken aback at the sheer size of it. Not only was the dog tall but he was broad with it, and had a thick dusky coat and pricked ears.

'Worra dog!' shouted one of the builders who was looking down from the scaffolding at the handsome and formidable creature.

'He's a Leonberger,' said Gustav, one of our incoming guests (or, as our first official visitor, a guinea pig depending on how you looked at it). 'Very friendly boy, aren't you, Hector.'

He patted the great dog on the head as it looked up adoringly at him. Hector's partially open mouth was upturned at the corners, giving the impression that he was smiling. Before we'd even begun our introductions to the rest of the party, Pippen had made her presence felt. She will square up to any dog that happens to be passing by and think nothing of curling back her lips and baring her teeth at the biggest, meanest-looking hound imaginable. Despite my protestations, and the rigid but dignified stance of Hector, she morphed from her usual docile, almost sleepy manner, into full attack mode. The children and I stood in the farmyard, mouths agape at Pippen's sudden show of viciousness. There was a hair-raising moment with Pippen springing into the air, her teeth gnashing and her ears flattened back. She didn't even come up to Hector's chest and the big dog stood bemusedly as Pippen raged away. When, finally, he tired of this ineffectual show of territorial dominance he just turned to the side and cocked his leg against the car wheel.

'Now, we are pet-friendly,' I reminded my incoming guests, 'but I must remind you that yer dog must be on a leash at all times when out and about.'

Hector was still wearing his over-animated dog grin and didn't look like a vicious sort, but looks can be deceiving.

Gustav nodded. 'Hector likes sheeeeps,' he said.

Hector was now straining at the leash, having glimpsed the peacock flouncing past.

I took a considered moment to think about what would happen if Hector ran amok amongst the sheep; it was a nightmarish thought. We had, in the past, had a few incidences of sheep worrying, and it was such a traumatic and senseless event that it really needed to be avoided at all costs. Poor Miles had even been through the trauma of having a dog rampage through his beloved flock of chickens. On that occasion, I had screamed at the dog owners, who stood rooted to the spot watching their black Labrador attack the hens. Their response, when I hollered that they should keep their dog on a lead, only served to compound my anger.

'But he is on the lead, I just let go of it . . .' came back the pathetic, whining reply, whilst feathers flew, chickens squawked, and Miles tearfully ran back and forth trying to shoo his hens away and grab the dog. I'd struggled to keep my composure and, in the end, after handing the dog owners an empty feed bag in which I demanded that they put the dead bodies, I had to just walk away with a sobbing Miles.

Afterwards, I was shaking like a leaf, a combination of anger and fright.

As a contract shepherdess some twenty years ago, I'd had the misfortune of witnessing the aftermath of a dog running loose amongst a herd of cows and calves, but this time the tables were turned and the dog came off worse. A cow's natural instinct is always going to be to protect her offspring from anything deemed a threat, whether that be a human or canine. We hadn't witnessed the incident but guessed that the poor dog had been crushed between a wall and a cow's head then stomped into the ground in a remarkable show of bovine strength from what usually are seen as the most mild and good-natured of animals. To lose a pet dog or farm animals in such violent circumstances is horrible for everyone and, in most cases, entirely avoidable.

Now, as I stood looking at Hector, I knew I had to make my point clear. 'Yes,' I said to Gustav, trying to sound nice and not too authoritarian, 'all dogs must be kept under control when in the countryside.'

'Right, you've seen Ravenseat,' I said to Gustav, 'so I'll take you to The Firs now. You'll 'ave to follow mi in t'car.'

I looked down towards the packhorse bridge and the picnic tables where I could see Pippen in full-on entertainment mode, executing a few cute moves in an endeavour to be rewarded with morsels of food. Chalky was coming back towards us, across the bridge and into view. She marched purposefully, her wiry brushlike tail erect, curled over at the end. She carried something in her mouth and, as I watched, the reactions of the walkers at the benches told me that all was not well.

I squinted, trying to identify the mysterious, elongated

object that protruded from her mouth at either side. Chalky stopped in her tracks, momentarily pausing for a readjustment. She placed her treasured item on the cobbled surface of the bridge then deftly picked it back up and proudly trotted up the farmyard.

'Your dog?' asked Gustav.

I nodded, then felt my cheeks colour as I realized that Chalky was carrying a sheep's leg in her mouth – and not just a part of its leg, the whole of it from hoof to shank end. Here was me telling people to put their dogs on a lead to avoid any sheep-worrying incidents, and my dog had just walked past with a severed limb in its mouth.

Clive appeared with a face like thunder, while Gustav seemed bemused and Chalky just looked mightily pleased with herself.

'Yer knaw t'yow that died o' staggers in t'Lang field, well I covered 'er up wi' a sheet an' put 'er out for t'knackers to pick up but summat 'as been at it,' Clive said, glaring at Chalky. 'An I knaw who the culprit is.'

'Aw Chalky . . . you haven't got a leg to stand on,' I said.

'Yes, she has,' muttered Clive.

Of course, Hector was never an ounce of bother. For all his fearsome size, his demeanour was that of a teddy bear.

I was very pleased when the builders finally left Ravenseat. For nearly a whole year, I'd had to put up with quips and comments coming from above; there was no hiding place. I could not retreat to the privacy of my house, as whichever window I looked out of there seemed to be a builder looking right back at me.

The extensive work undertaken was painstakingly tedious, day after day of picking out the old mortar and replacing it with new, but eventually the job was finished, and we were left in peace, cosy in our now-weatherproof and water-tight house. As the work on the farmhouse had progressed, we had started to feel guilty when we saw how much the roof of the cow barn was leaking when it rained.

'We'll 'ave to get it repaired afore winter,' Clive had said, 'but for now t'weather's set fine so we'll let 'em get calved and then lie 'em in.'

It was always best to let the cows stay outside for as long as possible if conditions allowed it. Our small herd of Beef Shorthorns were a native breed that could withstand harsher conditions but, once the ground became soft and the grass stopped growing, they would come inside until spring returned. They would spend their summers at the moor or in the allotments, roaming wherever their fancy took them. We would often lose sight of them for a day or two and then set out to go and find them. It was not a particular worry if they were all missing, but a real sign of trouble would be the absence of one of the tight-knit herd.

The bull ran with the herd in the winter, and we'd write when each cow was bulled in a diary, so we would have an estimated calf delivery date. When this time was upon us, the cows would be let back into the smaller pastures where they could be kept under closer observation. The Beck Stack or the Cow Pasture was the preferred field as it had shelter in the form of a barn and a small beck that never ran dry. It grew sweet grass and had undulations and gullies that afforded privacy to the mother-to-be who invariably wished to make herself scarce when in labour, her most vulnerable time and the only period that she really wished to be well away from the rest of the herd. After consulting the diary, it looked like we could expect our first happy event in September and, sure enough, that was when Eartha, one of our older beasts, began to show signs that she was getting near to her due date.

'Aye, she's warmin' noo,' I said, after I'd been up to feed the herd with a bale of year-over hay that had been earmarked for them. It consisted mostly of rushes and none of the other animals would have eaten it, but the cows would happily pick through it. Waste not, want not.

Her belly was distended, taut and stretched, her udder now full, though unevenly so. She had, in the mists of time, lost one of her quarters due to mastitis, but it proved to be of no hindrance to have only three out of her four teats producing milk. Any calf worth its salt would soon realize that when it latched on to the smaller, harder, front-passenger side teat it would not get a bellyful of milk, and frustration and pangs of hunger would soon send it looking for alternatives.

'It's nae sort coo that cannae rear a cawf on three cylinders,' Clive had wisely said.

Eartha had lumbered towards me and I could see she certainly hadn't lost her appetite, for she took a mouthful of hay and slowly chewed whilst studying me intently.

'It's time that we let yer through that gate and into't land where I can keep an eye on yer,' I muttered to myself.

Easier said than done, it was only going to happen with bribery, and a bale of rushy hay was not going to be anywhere near enough. I found more-edible temptation in the meal house – a bag of mollassed mixed cereals usually reserved for the tup shearlings going to sale. Clive and I went back to the allotment gate and looked at the cows some hundred yards away on a bare knoll, still hoovering up the last few wisps of hay.

'We'll let 'em all through't gate,' said Clive. 'It'll be easier than trying to separate Eartha from 'em.'

'Coosh lasses,' I shouted, then shook the feed bag. They put their heads up and stared intently – even Eartha seemed to have temporarily stopped chewing.

In this situation, one of two things might happen. In the ideal scenario, their inquisitive and greedy nature would bring them over and I could get them to follow me – and the food – quietly and calmly through the gate. The other possibility is that, rather than smell the molassed cereals, they smell a rat, panic and take off at high speed up the field. This time, greed got the better of them and the twelve cows ambled towards me.

'Coosh lasses,' I said again as I rattled the bag and stepped backwards towards the gate that would let them into the Beck Stack. They behaved impeccably, no rushing or barging at the gate, no one refusing to follow, they executed the move as peacefully as one could have hoped. I gave them their reward, a long line of feed on the grass, whilst Clive tied the gate shut, looping baler twine over the wall cheek topstone.

'Grand,' he said. 'Now we knaw where we 'ave 'em.'

Bovine surveillance now began in earnest. From the pack-horse bridge, an observer could see almost all of the field, count the cows and hopefully reach twelve. The children would count before school and again after school when they got off the bus. Secretly, I hoped that one day there would be thirteen and Eartha would have done what most native breed cattle do: give birth naturally to a healthy calf without any need for interference.

Alas, it was not to be. Four days later, Clive went to an evening meeting at Keld, and Miles and I set off to do the last check of the day before turning in. We stood on the bridge and counted the cows; all were present and correct, although one of them appeared to have distanced itself from the others and was up at the top of the field.

'Ach, look see, I bet it's Eartha, her time's comin',' I said to Miles.

We set off, chattering about nothing in particular. The exertion of the uphill walk meant that talk had dwindled to nothing by the time we reached the top wall where we'd last seen Eartha. It was a beautiful, breathlessly calm evening. The light was fading fast, the nights were now drawing in, and already there was a freshness in the air that filled your nostrils, cleared your head and sharpened your senses. We both stood quietly, surveying the scene. The rest of the herd were in below us, out of sight, but somewhere amongst these rushes and scrubby grass was Eartha.

I shivered, then gestured to Miles, who was crouched at the wall back, to be quiet. We followed the wall, Miles taking a slightly different line but parallel to me. He had only walked a few yards when he stopped in his tracks, waved and pointed towards a wide-open dry ditch. Standing quite still, staring fixedly ahead and right at us, was Eartha.

196

Her tail swished violently and rhythmically back and forth as she shifted her weight from one hind leg to the other.

'There's gonna be a calf soon,' I said to Miles.

Her belly twitched, a kick from within prompting her to turn her head around and look down at herself quizzically, as if to see where the movement had come from. She let out an audible sigh and then went back to chewing the cud.

The physical signs of an imminent calving were all there and, not wanting to interfere or upset what seemed to be the early stages of a straighforward labour, we left.

'How lang does ta think it'll be?' said Miles as we picked our way back down the hill towards home.

'An hour or two yet,' I said, since there was nothing to be seen under Eartha's tail, no sign of any hooves.

'Can I com' back wi' yer to check 'er a bit later?' asked Miles.

We agreed that when Dad came back from the meeting, we would all go and see that everything was going along smoothly. While Miles did his homework, I paced up and down like an expectant parent myself, looking at the clock, drinking tea and willing Clive to come back so that we could all set off and see that Eartha was all right.

'It's no good,' I announced, 'I cannae wait any langer.'

An hour and a half had passed, the light had all but gone, and there was no sign of Clive. I put on a head torch, Miles found a flashlight and we set off back to the field. This time we knew where we were heading, and we were feeling buoyant, confident even, that Eartha would be well into labour or ideally have already given birth to a healthy, strong calf.

The dismay I felt when we finally reached her was indescribable, for there she was, laid on her side, her head arched backwards and legs splayed but with absolutely nothing to show for all her straining.

'Christ Almighty,' I said.

'What we gonna do?' Miles asked.

And that was an excellent question because I wasn't quite sure myself. I decided that we should have a tentative examination to see if we could find the problem. The absence of any hooves was worrying, but if we could confirm that the calf was presented correctly then we would have to assume that it was a male and its size was the issue. Heifer calves, even as newborns, tend to be a little more delicate than the bulls. The other scenario was malpresentation, which could be anything from the unborn calf having a leg bent back to it being a breech, coming tail first.

Miles and I both quietly walked to Eartha's head. Her wide, unblinking eyes told us she clearly wasn't happy but the way she swung her head around as she saw us approaching illustrated that she was still feeling feisty.

'You're gonna 'ave to keep her laid for me Miley, mi lad,' I said.

''Ow am I gonna . . .' he started to protest, shining the flashlight on the heaving hummock of a cow.

'Like this,' I said, and in one deft move sprang towards Eartha and grabbed her nose tightly, my forefinger and thumb up either nostril. Holding her head sideways, keeping the pressure on, rendered her immobile. She could not rise with her head looking backwards. It was not strength that was required, just a stubborn refusal to let go.

Bless little Miles, he never faltered. Positioning the flashlight in a rush bob so that he could see what he was doing, he crouched beside me, one knee resting on the now-damp grass and the other gently on Eartha's neck, and got her nostril firmly gripped.

She snorted defiantly, and Miles looked disgustedly at the snot on his fingers.

''Urry up, Mam,' he said.

I nipped around to the back end. Lying down, I pushed her tail aside and felt around. There was a pair of front hooves just out of sight but no sign of a nose. Diving position is what you'd hope to find but, in this case, the fact that there was no sign of a head made me think that there was something wrong, perhaps the head was lolling to one side or perhaps there just wasn't enough room. At least we now knew what we were up against.

'Right, we need to ga an' ring t'vet, Miles.'

Miles released his grip and grabbed his torch. Eartha, I darcsay with a crick in her neck, sat up like a dog for a moment and then limbered up.

It was now a full-on emergency, and as Miles and I hurried back home, I was glad to see the lights of the Land Rover coming down the road.

'Whassup?' asked Clive, as I stood at the farmhouse door, bent over to catch my breath.

'Eartha's stuck a-calvin',' said Miles.

Clive launched into a tirade, ranting about how the moment your back was turned something disastrous would occur.

'Yah munt crib mi for it. I feel wretched enough as it is,'

I said, wishing to hell that I had realized earlier what was afoot.

To try and get the vet, and all the tools required, half a mile up a steep field in the darkness, and then have no lights or any way of restraining a cow, was going to make this job difficult if not impossible. We agreed that we should try and get Eartha into the barn in the field, at least then we could fasten her into one of the stalls with a halter and we could, in theory, take the quad bike, park it by the door and use its headlights to illuminate the dark recesses of the building.

Clive and I went to load the quad bike with things that we thought may prove useful – a halter, calving ropes and a small quantity of feed in the bottom of a bag – and Miles went to ring the vet, pleased to have been tasked with an important job that didn't involve a stressed-out cow. Within minutes, we were good to go. The older children had, as they always do, taken responsibility for the younger ones and were now busy showering, bathing and putting the babies to bed. Miles had spoken to Willy, the vet on call, and had given him a very brief rundown on the situation. Apparently, the conversation went like this:

Willy: 'Hello, emergency vet, how can I help?'

Miles: 'We've got a coo stuck a-calvin'.'

Willy: 'Where?'

Miles: 'Up in t'Beck Stack, Coo Pasture yonder.'

Willy: 'Who am I speaking to?'

Miles: 'Miles frae t'Ravenseat.'

Willy was already out on a call, a caesarian on a cow at a dairy farm near Appleby. He couldn't come straight away but he would ring when he had finished the operation to see if any progress had been made. Miles had to man the phone and await his call.

Clive and I set off back to our patient. We jumped off the quad bike and I strode confidently through the rushes towards where Eartha had last been seen.

'Oh, great,' I muttered, 'she's gone.'

It might seem like the simplest thing in the world to be able to find a substantially sized – and I mean a half ton – cow in a field, but what followed was more than four hours of hunting high and low. We searched methodically, starting where Miles and I had left her.

The night sky was clear but, though the moon cast a little light over the terrain, it was still ridiculously difficult to keep our bearings. I switched my head torch off to see whether my eyes would adjust to the blackness but, even after a minute or two of standing quietly, I could pick out little more than an outline of the drystone wall on the horizon.

Clive had set off to the farthest corner of the field and I could see the beam of his flashlight sweeping from side to side as he searched amongst the rushes. I stumbled and tripped my way along, the head torch back on. Occasionally, I'd put up a pheasant or partridge, and my heart would miss a beat at the resulting squawk and flapping of wings as the bird disappeared off into the darkness. My ears tried desperately to pick up any kind of noise that might reveal Eartha's whereabouts. The presence of the other eleven cows, as well as a handful of sheep, didn't help matters. To see in the torchlight the reflection of a pair of green eyes staring nonchalantly back at me would momentarily fill me with hope that I'd found my missing cow but, in the end, we accounted for every single member of the herd apart from Eartha. As soon as my presence was deemed to be of no significance, the cows would look away and resume their cudding while the sheep would turn tail and scoot away into the undergrowth.

'Anything?' shouted an increasingly exasperated Clive.

'Nowt so far,' I shouted back.

We decided that the next course of action should be to use the quad bike to search the flatter bottom-half of the field. What with the bike headlights beaming straight ahead and me riding shotgun, shining the torch to the side, we could cover the ground faster. We did a couple of sweeps, didn't locate Eartha, but did find every single rut in the field. We jolted and bounced along, Clive cussing under his breath whilst I held on for dear life.

After that we decided to return to the farmhouse to give Willy the somewhat embarrassing message that our labouring cow was missing and that we would ring him when we had found, and apprehended, the patient. Back we went, through the gate, and homeward bound.

I made the call to Willy whilst Clive made tea. It was nearly midnight and the bigger children were still awake and eager to come and help. I promised that I would get them when we found Eartha but, for the moment, I needed them to stay put and watch over their sleeping siblings.

'I can't believe it,' I muttered in between sips of tea. 'I knaw that field like t'back of mi hand an' yet in t'dark I'm completely lost.'

We agreed that we would return for one more attempt,

taking a feed bag and giving it a shake to see if the promise of food would tempt Eartha out of hiding. Personally, I didn't think that food would be at the forefront of her mind, but these were desperate times. Eleven cows thundered towards us out of the darkness, bawling and snorting in anticipation of an impromptu midnight feast. We conceded defeat and, in a huff, left the jostling herd of cows to eat up their ration. Clive said that we should sleep for a couple of hours and, at dawn, the rising sun would throw light onto the situation.

'Ya dun't think that she's dead?' I said.

'No,' he replied, 'but likely t'cawf is.'

I began apologizing, admitting I'd made a grave mistake in leaving her be. I had always been a big believer in letting nature take its course, and not intervening unless really necessary, but this time I had made the wrong call and it hurt.

'Nae, it cannae be helped,' Clive said, turning to me.

Back we both went for another sweep of the field. When driving the quad bike it isn't a good idea to take your eyes off the way ahead, especially when navigating your way across an uneven field in the pitch black. A narrow deep gutter cut through the field – and we went straight in it. It wasn't big enough or deep enough to cause either us or the quad bike any lasting damage, but I was catapulted forwards and landed awkwardly on the ground at the other side. Clive was still astride the bike, though now sitting on its tank. Its rear wheels were stuck up in the air, the engine still running.

An exchange of expletives followed.

I brushed myself down and surveyed the damage. My left foot hurt, quite a lot actually. Clive was physically all right but not best suited.

'Gis a hand to get t'bike out,' he said gruffly.

I limped over.

'Stand at t'side and press t'throttle,' he said, warning me that I'd have to be ready to move out of the way quickly for, if the bike couldn't grip sufficiently up the incline of the bank, there was a danger it might roll over. I pressed the throttle gently and the bike moved forwards, the back wheels dropping down and the front ones rising up at the other side. The front headlights that had temporarily illuminated the muddy water that ran through the gutter now shone upwards and onto an eroded shingle cliff, upon which a lone mountain ash grew. There, below the tree, stood Eartha!

'Would you believe it?' Clive muttered. 'After all this flamin' time. What the 'ell is she doing there?'

'Hiding . . . in plain sight,' I replied.

We agreed to abandon the bike, knowing Reuben would take great delight in rescuing it the next day. Our priority was to get Eartha back to the farm and into the barn, where we would have sufficient lighting and be able to keep everything clean, as it was highly likely that a caesarian would be required.

'Are yer gaan to open t'gate or are yer gaan to fetch t'coo?' Clive asked.

I had a feeling that my gentler manner might prove more successful at persuading Eartha to walk home. Clive seemed to be exuding some pretty cross vibes.

'You do t'gate,' I replied, already limping off in a wide

arc so as to get around the back of Eartha in case she had any notions of heading for the hills.

'Cooosh lass,' I said.

She looked towards me, then took a few tentative steps down from the cliff. She seemed to have resigned herself to the idea that she needed help, as there was no more cavorting or snorts of disapproval as she quietly trudged towards the distant lights of home. Hobbling alongside, I squinted, trying to see whether there were any signs of the calf underneath her tail as it swung from side to side.

Clive walked ahead, opening the gates and making clear the way, across the packhorse bridge and between the barns. Eartha never put a foot wrong and it grieved me to think of what lay ahead: an operation, a dead calf and then a long recuperation. I felt very responsible and vexed at making the wrong decision.

It was 1.45 a.m. when Willy appeared in the yard. He'd had a busy night and had just dozed in the chair by the fire in between callouts.

'Have yer had a pull?' he enquired, as he scrubbed up.

Clive and I explained the circumstances and that we had not attempted to calve her; the calf's front hooves had been felt but with no sign of a nose, so we had left it well alone.

'Good, good,' Willy said.

That was the first time I felt that I'd done anything right that evening.

Raven, Reuben and Miles appeared, Willy's car lights coming up into the farmyard having woken them up. Eartha was behaving impeccably, standing stock-still while Willy examined her.

'Ach, it's a big enough calf,' he said as he rested his chin on Eartha's flank and I held her tail aside to prevent him being whipped in the face with it. The children sat upon a straw bale in dressing gowns and wellies, watching intently.

'I think that we need to do a section,' he said, withdrawing his plastic-gloved hand.

'Oh,' I said, 'really?'

I had, in the past, seen an unborn dead calf or lamb dissected internally and removed in pieces. An embryotomy was a horrible undertaking, but if the calf was already dead then it did save the poor mother from the added trauma of a caesarian.

Raven, bless her, always blunt, asked what I was thinking: 'Is t'calf dead?' she said, her feet swinging, and her dressing-gown collar pulled up to beneath her chin, her long shiny red hair tumbling down to her waist.

Miles and Reuben looked on wide-eyed.

'Good God, no!' Willy exclaimed. 'I felt his hoof twitch just then.'

'Let's be gettin' on then,' said Clive.

Eartha was haltered and prepared, shaved and daubed with antiseptic where the incision in her side was to be made, and given a local anaesthetic. A little time was needed for the anaesthetic to take effect and the area to become numb, during which time I scrubbed up. An assistant is helpful when it comes to delivering the calf and, afterwards, during suturing.

Clive held the halter, as sometimes the cow would lie down and, if it was tied securely to a gate or tethering ring, there was a possibility that it could strangle itself. Eartha did indeed lie down, and Clive was able to push her the right way, so Willy could make the cut at the prepared site without hindrance. So far, so good. At one caesarian I'd attended, many years ago, the cow decided to lie down half way though the operation, which didn't make for the most sterile of surgeries. As the cow collapsed onto its side into a bed of straw, a cloud of dust and detritus filled the air and gently fell all over the gaping wound as the vet sprawled over her side, trying to act as a human shield. It was a miracle that the cow didn't get peritonitis or septicaemia.

The whole procedure took an hour. Raven watched intently from the top of the straw bale, asking Willy the occasional question and eventually persuading him to give her a week's work experience during the next school holidays. Reuben and Miles milled about chattering to each other; Clive and myself said little, just watched, waited and hoped.

The arrival of the calf was met with stunned silence from the children. The operation, in the hands of an expert, seemed so simple but the miracle of life itself was there to see in front of their very eyes. Without ceremony, Willy pulled and lifted the calf from Eartha's womb. The sheer strength required to both hold and guide the calf's torso through the incision in the mother's side was incredible.

'You hold the back legs and keep it coming, Amanda,' he panted as the calf emerged. 'Clive and Raven, work on the calf, Dopram-V drops are on the side.'

Clive grasped the calf by its front legs and dragged him out of the way.

'Amanda, we need to stay clean and close,' Willy said, seeing that I was now distracted by the motionless calf.

Raven was unscrewing the small bottle of dopdrops. Sometimes, it could seem to take forever before a calf or lamb would breathe, particularly after a traumatic birth. A drop or two of this substance, which is a breathing stimulant, would often result in the animal taking its first, very sharp, intake of breath. Inside, I was willing Raven to hurry. The calf had been stuck for long enough that his head had swollen, his tongue also.

Clive had cleared the mucus from the calf's airways and was now slapping his wet chest with the flat of his hand. Those moments, which, in reality, were probably only seconds, seemed to go on forever, but eventually Clive held the calf's mouth open whilst Raven dispensed far more drops than were recommended onto his tongue. The effect was instantaneous: the calf gasped, threw back his head and then brought it forward in an almighty sneeze. He then lay panting, almost hyperventilating.

Willy turned around, looking over his shoulder towards the wheezing calf behind him, and he gasped too.

''Ow much did you give him, Raven?'

'Erm,' she mumbled, holding the bottle up towards the light. It looked empty.

'Anyway, it looks like he's a runner,' Clive said, changing the subject.

'Close run thing, though, mind,' said Willy, who was now sewing Eartha back up. I held the suture knots taut with forceps whilst he stitched together layers of uterus, muscle and eventually skin. 'I'd get some colostrum off her while she's down.'

That first feed of colostrum, full of antibodies, fat and

nutrients, is essential for a newborn. It gives them warmth, energy and kickstarts their immune system. Eartha needed a few hours to recover from her ordeal, so we were temporarily tasked with feeding the calf from a bottle.

We thanked Willy for his efforts as he swilled off, cleaned up and packed up his equipment. A better outcome we could not have wished for. I took one last look over the barn door; Eartha remained in a stupor, shellshocked, sitting with her head upright and staring blankly at the wall. Her calf, a beautiful roan bull, was lying near her, protected by a small wall of straw bales so she could see him but could not accidentally roll on him before morning.

'We'll leave the building light on til't mornin',' said Clive. 'Then if she 'asn't risen we'll 'ave a go at gettin' her to her feet.'

The children had returned to their beds to grab just a few hours' sleep before school time, though, as the night sky did seem to have lost some of its inky blackness, I feared that dawn was not so far away. It had been an eventful night and the adrenalin coursing through my veins had taken my mind off my injured foot but now, as order was restored, unfortunately the pain returned. I sat down on the bench outside the kitchen door and looked at my wellies. They certainly needed to come off before I ventured into the house. The right one I kicked off without issue but the left one was going to prove problematic, feeling overly tight on my throbbing and seemingly swollen foot.

'Want a pull?' asked Clive.

'I dunno, Clive.' I winced, thinking about the force that was going to be required.

'It's gotta come off, Mand,' he said.

The possibility of cutting it off did flash through my mind, but it seemed like a waste of a good welly. Clive was thinking the same.

'Should 'ave asked Willy to remove it wi' t'scalpel,' he said.

I grimaced.

'Mand, you've 'ad nine babies for Christ's sake, one on yer own by t'fire. Yer as tough as old boots . . . haha . . . get it?'

This was his standard response now to any kind of ailment or injury I suffered whilst going about my duties.

I capitulated, and Clive pulled it off. The pain was excruciating; however angry my foot was before the welly came off, it was now multiplied by a thousand.

'Better?' asked Clive.

I cannot recall my answer, but it was probably unladylike.

I took off my socks and studied my left foot. It looked particularly ugly – red, blotchy, puffy and considerably bigger than my right one.

'C'mon, Mand, let's go to bed,' he said. 'We'll 'ave a look at it in t'morning.'

I reminded him that it was the morning, but it fell on deaf ears as he'd already left the room.

I didn't get a wink's sleep, just the bedsheet touching my foot was enough to cause pain and, by the time I was giving the unwelcome wake-up call to the oldest three children, I had a foot that resembled a mottled red-and-purple rugby ball.

'Do you think yer should go an' 'ave it looked at?' asked Clive.

I shrugged; the mere logistics of organizing a trip to Accident and Emergency were unthinkable. Clive would have to drive as I couldn't bear to use my foot on the clutch, Annas, Clemmie and Nancy would need to come with us, we'd need to find someone to be at home for when the rest of the children came home from school. Then there was the farmwork to contend with, not forgetting Eartha and her calf to see to. That was before even considering who was going to welcome the guests that had booked into the shepherd's hut. Just thinking about it all brought me out in a cold sweat.

'It'll be fine, honestly,' I said.

'Twiddle your toes,' said Clive, the classic test for broken bones.

The unofficial diagnosis from various members of the family and friends was that it was just a badly bruised foot and that it would right itself eventually. I hobbled about the place wearing one welly and just a bare foot, using an unclaimed NHS crutch that someone had once left at the farm.

The following day, I spent a fair bit of time sitting on a straw bale watching Eartha with her calf. Clive had got her up and she had pottered around the barn a little bit, no doubt still sore from the whole episode. She had bonded with her calf; he too had now found his feet and latterly the milk supply, her udder.

Sometimes luck is on your side and, on this occasion, it certainly was. Only days later Eartha was back outside in the garth cropping the grass with her calf by her side. My foot trouble was minor in comparison with what might have happened that night; a miss is a miss, they say, there are so many 'what ifs' and 'maybes' in life. A minute or two later and the calf might have died, sooner and perhaps the caesarian could have been avoided and the calf born naturally with the vet's assistance. There is no use speculating, you just take whatever life throws at you and deal with it however you see fit.

It was whilst I was hobbling with the crutch that I was asked to go on live TV to talk about the issues of living rurally and, in particular, the lack of public transport in remoter regions of the country.

'I'd love to do it,' I said to the researcher on the phone, 'but when?'

'Short notice, I'm afraid,' she replied. 'Day after tomorrow.'

'Where's it at?'

'Studio,' she said. 'Salford.'

'*Salford?*' I exclaimed.

'Tell 'er you've got a gammy leg,' piped up Clive who was studying a flock book but really earwigging.

'I don't think that I can make it, I'm afraid. I's lame, I can't drive at the moment,' I said and then went on to explain the whole sorry story.

'Don't worry about that,' she said. 'We'll get you there.'

All I had to do was await the arrival of a taxi to take me to the train station, then a car would pick me up off the train and get me to the studio. It was a perfect plan . . . apart from the bit where the taxi driver, who was booked from a rank thirty miles away, didn't allow enough time to get to Ravenseat and then the train station. The opportunity to go and talk about the problems with rural transport was thwarted by rural transport itself. The irony.

Clive found it very amusing.

'So, you didn't ga on telly to talk about the trouble wi' transport 'cos yer missed the train. A lame excuse if you ask me.'

7

The Beast from the East

'No, it's not acceptable to mend yer motorbike in t'kitchen,' I'd said to Reuben one evening after a visit to a friend's farm where some kind of operation was being performed on a trials bike in the kitchen. Oily parts had been strewn across the newspaper-covered kitchen table.

But it is around these kitchen tables that plans are made, gossip exchanged, and tales told. Every weekday at dinner-time, the postman, after a cursory knock, opens our door and tosses the day's post onto the table. Everyone stops eating and drinking for a moment while we study the envelopes or parcels.

'Lamb cheque,' gripes Clive as he tears open the envelope and looks solemnly at the sheet. 'Trade was nowt, lamb prices were back last week.'

'*Northern Farmer* paper,' says Reuben, unfurling it and skipping straight to the back section with the machinery sales.

On one occasion, there was a horrified squeal as he caught sight of a picture of myself in there.

'Gawd, I cannae get away frae yer,' he muttered, 'yer everywhere. What's this, you're epileptic? Aw din't knaw that.'

'Eclectic,' I said.

One mysterious, long rectangular package arrived at Christmas and, to the great amusement of the curious onlookers, was emblazoned with an amusing logo.

'Horny Hobbies,' read out Reuben slowly.

'HORNBY,' I corrected him quickly. 'HORNBY hobbies, it's a model train, for goodness' sake!'

I opened an unexpected parcel that arrived one lunchtime.

'A gift,' I said, after reading the label, 'from someone who stayed in t'shepherd's hut a few weeks back.'

'I bet it's a knitted sheep,' said Clive wryly.

'No,' I said, smiling because I knew the reaction I was going to get. 'It's from a gentleman an' it's fishnets, actually!'

'Whaaat? The cheeky so-and-so!' retorted Clive.

'Yes, apparently if I bait them up with a piece of sausage or bacon, I'll catch forever of trout in t'beck.'

No two days were ever the same at Ravenseat, and quite who would turn up nobody ever could predict but there was one thing for sure, there was never a quiet moment. It took a downfall of snow to keep the visitors at bay.

When the children complained that it was a hassle to have to walk from the road end down to the farm when it snowed, and the school bus could not travel the last mile on the untreated road, I'd remind them that in the past nearly all journeys were undertaken on foot. On one such occasion, a blizzard hit and an unfortunate traveller, Henry Wastell, was lost. He was found dead three weeks later, standing bolt upright in a peat hagg at the top of Tailbrigg on the road to Cumbria. An observer at the time remarked that the frozen body was a thoroughly terrifying sight to behold.

When the weather turns, it can do so with a vengeance, you only had to look at Muker's parish records to find examples of people who had been going about their daily business only to succumb to the elements. Essabel Scaife 'perished and dyed' as a result of the tempestuousness of the weather on 23 November 1641. Drownings, too, seemed more commonplace, as folks took shortcuts on their way back from working in the mines or labouring in the fields, although on Friday 4 July 1824, John Harker lost his life whilst bathing in the River Swale.

In *Swaledale*, a book written by Ella Pontefract and published in 1934, she writes about talking to two old men at Ravenseat about the hard winters they used to have up here. 'We doa't hev winters like them noo,' they'd said. But only a few years later, in 1947, there was a terrible winter when the dale was completely cut off, and men were digging their way out of the drifts that engulfed the cottages in Keld. There have been bad winters since, and the heavy snows in March 2018 were, as I was constantly told, reminiscent of the storms of those times past. A deadly combination of howling gales, biting cold and driving snow under which walls disappeared, rivers froze, and roads became impassable. Nicknamed 'the beast from the east' because it originated in Siberia, it was a storm of intense severity that wreaked havoc across the country. Schools were closed, power was lost, and, in some areas, helicopters were used to drop food supplies for humans and animals in stricken communities cut off by the snow.

The problem for many farmers during the storm – ourselves included – was not having enough feed for the animals. Good quality forage, hay and silage, was in very short supply after the washout of a summer in 2017. Our saving grace was that this blast of wintery weather was exceptionally well forecast. In the frozen days before the storm hit, we filled the proven store with sheep cake, the barn with fodder beets, and the dairy with milk, cheese and other perishables. It didn't matter what the weather did, we were prepared for anything. Getting a fifteen-ton load of fodder beets in had been pricey, but the peace of mind it gave us that we could feed all the hungry mouths, of which there were some eight hundred, was priceless. We'd weaned the sheep onto them over the past few years and now they ate them heartily. The cows were addicted to them and would bellow for them at breakfast time and tea time and then crunch their way through them noisily.

The fodder beets were grown around the Scotch Corner area, some thirty miles away, so transporting them to Ravenseat by tractor and trailer was a costly and time-consuming business. After stopping off at a weigh bridge to calculate the weight of the load and thus the price, they would be brought up Swaledale, no doubt to the annoyance of any cars stuck behind the slow-moving rig.

It always took a bit of careful manoeuvring to get the fodder beets unloaded into the protective straw compound which we had constructed to prevent them from getting frost-damaged. We would watch for Edward, our feed haulier's, arrival and, as soon as we saw the flashing neon light approaching, would chase the children indoors to safety and start moving around all of the general detritus that lay in the farmyard in an effort to get the tractor and trailer backed in carefully.

'Edward's comin' down t'road,' Clive had said that winter. 'Skidaddle them kiddies into t'ouse.'

They were happy to go in and thaw out, for there was an icy wind already blowing, the precursor of what was to come. They sat on the sill in the living room, peeping through the window and occasionally giving me the thumbs-up sign.

The mile-and-a-quarter single-track road into the farm is only partially visible from the farmyard, but the sheer size of the tractor and trailer meant that its orange flashing light could be followed as it steadily made its way towards us. Clive and I scurried off to make our preparations; I moved the Land Rover, Clive shifted the wheelbarrow and picked up the children's scooters.

We regrouped in the yard, having made enough room to afford Edward free passage to the barn, but it seemed that the flashing light had stopped moving some half a mile away. Reuben and Miles appeared from nowhere and joined Clive and me in staring up at the road awaiting the tractor's arrival.

''T in't movin',' said Miles, stating the obvious.

'Summat must be wrang,' said a disgruntled Clive.

It was. Edward had only one more steep hill to negotiate before reaching Ravenseat but, as he'd lightly applied the brakes before the descent, he'd skidded on some black ice and come off the road. The tractor was still facing in the right direction, but the trailer had slewed across the road, its two nearside wheels sunk up to the axles in the soft roadside verge. The trailer was now tipped sideways and the tractor immovable. Edward was mortified; he'd travelled for thirty miles at a snail's pace without incident, and now, within sight of the drop-off point, he'd come a cropper.

Big and broad-shouldered, ruddy-faced and with a thatch of auburn hair, he was a man of few words but a great thinker. As sharp as a needle, he could convert kilos to tons, acres into hectares, talk about crop yields and work out dilution rates in the blink of an eye, but, unfortunately, in this case there was to be no mathematical solution. Problem-solving might have been his forte but the answer to this setback was simple. He stood, hands on hips, and surveyed the scene.

'There's only one thing for it,' he said stoically. 'Them beets, we gonna 'ave to handball 'em off.'

The trailer would have to be lightened before the tractor could pull it out. Miles sighed loudly.

'Can't we unload 'em wi' our tractor and loader bucket?' asked Reuben hopefully.

That would have been the labour-saving way to do it, to open the back door and scoop them out, but the hydraulic trailer door was heavy enough that to raise it entirely endangered the stability of the whole trailer. We gave it a great deal of thought as we went to get our tractor and another trailer, but we came back to the same conclusion. For safety's sake, the fodder beets needed to be unloaded by hand.

What a day that turned out to be! We took it in turns,

forming a human chain, little ones rolling the beets forward whilst bigger children threw them onto our trailer. In lighter moments, the jokes came thick and fast.

'Yer knaw what they say . . . if you cannae beat em . . .' said Clive.

We all groaned.

'Going back to mi family roots,' I said, evoking a similar response.

The frustrating thing was that the pile of beets never seemed to diminish. The fact that they were roughly spherical meant that as quickly as you moved them, others from the considerable heap rolled into their place. Eventually, hours later, when we had shifted the majority of the load and were thoroughly fed up of the sight of fodder beets, Edward decided to have a go at freeing up the trailer.

'Time for Edward to beet a hasty retreat, dun't you think?' said Clive, trying to raise a laugh.

We all nodded. There are times when reverting back to the old-fashioned way of doing things does not suit, and this had definitely been one of those occasions.

The snow arrived as predicted on 27 February. That morning, the sky took on a heavy, leaden appearance. The cold wind that had blown for the previous few days calmed, an almost eerie quiet descended, and the landscape took on a singularly bleak, sterile appearance. You could smell the impending storm in the air. The children didn't go to school; there was no reason to risk them being trapped there, unable to get home.

We brought down the moor sheep as the flurries of wintery flakes turned into a driving mass of blinding snow. I drove the quad bike headlong into the blizzard, whistling for the sheep. Miles sat on the back, entirely covered from head to foot in layers of clothes, only his eyes visible from under a balaclava that was topped off with a woolly hat. He ducked down, sheltering behind me as I drove upwards and out through the moor gate. It was intolerably raw. My knitted hat was unfurled and pulled down to flop over my eyes. The wool was heavy with a layer of crisp solid ice crystals that left my forehead beneath numb. I was breathing through a cotton scarf that was pulled over my mouth and nose; it was soon wet on the inside through condensation.

Stopping the bike, I got off to walk to the ruins of the Robert's Seat watch house. I was unable to look up into the maelstrom of snow being whipped up, but on foot I could walk backwards and occasionally glance over my shoulder to see where I was going. Miles came along too, walking with his back to the howling easterly gale, his chin buried in his coat and head dipped down as though looking at his feet. We kept close but didn't speak to each other, it was trouble enough just to breathe. Even Kate, my sheepdog, with a never-ending source of energy, could not keep her face to the weather. She would run for a while, then turn tail to the wind and scrape her snow-smattered face repeatedly up and down her front legs in an attempt to defrost her encrusted brows.

Robert's Seat was the highest point of the moor and where we had been feeding the sheep daily prior to the bad forecast. In normal conditions it was a great vantage point commanding an unparalleled view across the broad acres of moorland upon which the Ravenseat heaf of sheep roamed.

That day, visibility was poor, down to just a few feet in front of you. I whistled, the piercing noise partially drowned out by the tumultuous roaring wind and muffled by the snow that lashed down upon us. It was the most extreme conditions that I had ever encountered, it was debatable as to whether we should really have ventured out into the eye of the storm, but if there was to be no let-up and the blizzard continued then there was every chance that the flocks would be buried.

We stood beside the ruined building, crouching down to gain a moment's respite from the bitter weather and to assess the situation. Out of the swirling white void that surrounded us, came sheep. There was no depth to the snow as yet, so they had no trouble with mobility, but they seemed to have taken leave of their senses and were disorientated, their vision impaired by the snow that clung to their faces. Kate had gone to gather them but there was little point in me trying to communicate any commands for I could not see either her position or indeed any of the sheep that needed guiding towards home. I would have to hope that her natural instinct to bring the sheep in my direction would suffice.

I leaned in close towards Miles.

'You alreet?' I asked. I could see in his eyes that he was cold, frozen, but probably, like myself, the adrenalin was pumping, and he felt exhilarated, alive. He nodded vigorously.

'C'mon then, we need to be ga'an, movin' towards hame,' I whispered, pressing my mouth up to where his lug lurked beneath the hat. I grabbed his arm and, raising ourselves to our feet, we turned through three hundred and sixty degrees. A sea of sheep besieged us, the only soupçon of colour amongst the multitude of whites and greys that made up the flock was the crimson mark that distinguished them as being from the Ravenseat heaf. We couldn't count them, it was impossible to even hold your gaze in one direction for more than a few seconds before you had to raise your hands to shield your face from the brutal onslaught of blizzarding snow. I signalled to Miles that we should leave. Kate had reappeared and was now standing resolute and determined at the back of the flock. With her fur blown backwards and snow balled up beneath her belly, she appeared to be twice the size of her normally wiry frame.

We set off back, arm in arm, downhill this time, the sheep, all tightly flocked together, revolving around us. Occasionally the tight-knit group would almost swamp us, and individual sheep that had been temporarily blinded would run into the backs of our legs almost knocking us to the ground. It was a blessed relief when we reached the bike and could get ourselves and our entourage moving quickly down to safety.

We crossed the beck, the sheep following obediently, and finally we had the flock in the sheep pens. Kate went back to her kennel to thaw out whilst Miles and I congratulated ourselves on a job well done.

'Let's have tea an' warm up,' I said, 'while Dad ga'as for t'next heaf.'

'Aw think we should give Kate a treat,' said Miles. 'She's done good.'

He was right, it was sometimes easy to overlook the loyalty of our sheepdogs. Never sick or sorry, they would always be on standby at the kennel doors desperate to accompany us and head out into the worst of weather. They took their duties seriously.

There were many stories of sheepdogs and their bravery and unfaltering dedication – or maybe dogged determination would be the most apt phrase. Like a good horse, the name of a good sheepdog would live on and their feats of endurance would be talked about for decades.

''Ave I ever told yer about Ken?' I said, gripping a mug of steaming tea in the kitchen.

'Joiner Ken?' asked Miles, leaning against the kitchen range.

I shook my head.

'Farmer Ken, from Thwaite yonder?'

'Nooo, Ken the dog,' I said, rolling my eyes.

The tale is set during the great storm of 1947 when it snowed, snowed and then snowed some more. The shepherd whose name hasn't gone down in the annals of history but was from the locality had set out early one morning, just as we had, to bring his flock down from the moor during one of the blizzards. He had just reached his flock when there was a break in the storm, visibility had improved, and four sheep were spied in the distance. The shepherd had sent Ken to go and round them up and bring them back towards the moor gate. Ken set off away into the distance – a decent fell dog will think nothing of taking the widest of arcs in order to get to the back of the sheep before they have a chance to notice and make a break for freedom. Then the snow began to fall heavily again.

'Stowerin', it was, Miley,' I said. 'All's white, a complete hap up, could see nowt.'

Miles nodded, we'd just walked right out of exactly that same situation.

The shepherd and the main flock of sheep walked back down off the moor top, all the while shouting and whistling for Ken, but his calls were drowned out by the raging winds and he duly accepted that, for the moment at least, he was alone without his faithful colleague. Ken was missing in action, as were the four sheep. It was after darkness fell that the shepherd began to worry for his dog's safety and decided that he would take a lantern and retrace his steps back to the moor gate in the hope of finding him. He trudged for an hour through the snow until he got to the moor gate where the light from his lantern reflected eyes looking back at him, four pairs belonging to the sheep that he had originally sent Ken to rescue. Beside the sheep lay Ken, just his head showing, his body covered with a foot of snow. He had done his job, carried out his instructions and gone above and beyond his duty by bringing the sheep to safety and standing guard over them.

Miles had tears in his eyes, I swear he did. He said that he hadn't and that it was just that being out in the cold wind had made his eyes water.

'Ken weren't dead,' I added. 'Shepherd carried 'im 'ome an' 'e made a full recovery.'

'I'm gonna cook up some warm gruel for t'dogs,' said Miles. 'Is that all right?'

It was absolutely fine. On days such as these the sheepdogs were what made our tasks doable.

Clive and Raven went for the Birkdale Common heaf of sheep. These sheep lived on perhaps the most dangerous ground of all, spread as they were over such a wide area that we would stand no chance of finding stricken overblown animals if they were buried under the snow. They would have to come right down from the moor and into the West End fields, where there stood one of our bigger stone barns that could house them all should the weather not improve.

The doors were thrown open and the hayracks inside filled with our precious meadow hay, of which we had so little. It was at times such as this that the little bales really were worth their weight in gold because they could be carried on our backs to the outlying places. The fodder beets were also most welcome, though we could not deliver them to all of the sheep as by the time it had finished snowing, two days later, the fields were inaccessible other than on foot.

As the blizzard raged on into its second day, Clive and I set out to deliver rations of the concentrated sheep cake to all the sheep who, although now safely down off the moor-tops, were mightily hungry. With not a blade of grass in sight, they were entirely reliant on us bringing them their nourishment. Two bags of cake, each three-quarters filled, were tied at the top with baler twine, ready to be carried. This mixture of cereals and molasses would be compensation for the lack of hay. I took one bag, Clive the other. We each hitched a bulging bag onto our shoulders and, after a bit of readjustment, we got them roughly balanced and set off. It was slow going. In some places the snow was barely covering the frozen earth but in other areas we were wading knee-deep. The flock were in sight when it all became too much.

'I can't do this anymore,' I said, leaning forward and letting the heavy bag drop onto the snow. 'I'm freezin'. Mi hands is agony 'cos of mi split finger ends an' I can hardly move 'cos of all mi layers o' cleaths. I'm knackered.'

'Naw I know,' said Clive, depositing his bag onto the snow too.

We commiserated with each other for a few minutes whilst taking a breather, then set off again. The snow had drifted and been sculpted by the winds into awe-inspiring perfect waves that followed the lengths of the walls, the crests of snow glistening in the sun.

We reached the sheep and they seemed most grateful, following us around as we put out little piles of cake here and there in the vicinity of the building.

'We'll be back wi' a lal' bit o' hay later, lasses,' shouted Clive as we set off back to the farm and then the next flock of sheep.

The barns served their purpose, keeping the weather at bay. The slitted windows allowed a minimal amount of light inside and, although the majority of the sheep did venture into the barn to nibble at the hay, plenty of them refused point blank, choosing to keep out of the wind at the back of the barn rather than take shelter inside.

One heaf of sheep had been moved into the Big Seeds pasture, which also had a barn that they could use for shelter. The sheep were physically nearer the farmstead but getting to them involved crossing the valley bottom where the snow just kept piling up. Commando crawling across snowdrifts that were waist deep at their shallowest was insanely fatiguing. Shimmying on our bellies, resting our weight on the full feed bags whilst pushing them in front of us must have looked particularly graceless, but we were left with no choice.

Reuben and Miles were tasked with thawing out water troughs in the cowshed after the pipes froze.

'Fourteen kettles so far, Mam,' said Miles flatly. 'An' the trough's still frozen. I hate snow.'

The smaller ones were banned from venturing outside at all, it was just too dangerous, and as long as the snow kept falling, they were assigned the very important job of feeding the fire and keeping the house warm.

The horses, who were used to being turned out every day, were restless in the stables. They were happy enough to stand with their mangers brimming with chaff, a mixture of chopped hay and straw that I could buy at the feed store. It was a good substitute for the precious hay and meant that I could avoid any arguments with Clive over the value of horses, seen as an expensive hobby compared to sheep, which were deemed profitable. But for all the food put in front of them they still took badly to being incarcerated. The sound of stamping hooves and snorts of disgust emanating from the building wore me down, until finally I capitulated.

'Reet, yer wanna ga outside,' I said crossly as Josie pulled yet another excruciating face when I went in the stable to poo-pick. Princess, a more sober character, stopped chewing and hung her quivering bottom lip. Her dark eyes peeped out from behind a thick forelock that was strewn with chaff and wisps of tangled straw bedding.

229

Josie paced to and fro restlessly. She was an uneasy horse, mainly of good temperament as her mother Meg had been, but she could be irritable on occasion. She was now the matriarch of the herd and with that came the responsibility of keeping the others in check. She would scowl and shove her way about, jostle for position with Della the bay, and Princess the piebald, and even back up to Little Joe and give him half-hearted kicks with both back hooves.

I opened the stable door, slipped off Josie's halter and watched as she manoeuvred her ample apple-shaped behind through the opening. Princess looked on anxiously until I unclipped her from the stall. After grabbing another mouthful of chaff, she too limbered off, though with no sense of urgency.

'Let Della and Lal' Joe out,' I shouted to Edith and Violet. 'Yer might 'ave to dig out the door a bit.'

They were in the farmyard messing about with what they optimistically called a snowboard. It was one of Reuben's car-boot sale purchases and the fact that it had a rudder attached to the underside made me think that it had actually started out life as a waterskiing board.

I carried on tidying the stable, sweeping the floor and banking the clean reusable straw up against the walls. The wet, dirty bedding would go into the rubber skep, a low-sided wide bucket that I could carry to the midden. The wheelbarrow was not going to travel through this depth of snow. I was in my own little world, thinking of all the horses I'd ridden, looked after and loved. Meg, Queenie, Scattercash, Beau, Bruno, Ember and Stan. I smiled, being in the stables was a tonic. I loved the smell of the horses; to push their manes aside and bury my nose into a horse's neck and breathe in the scent was therapeutic and good for the soul.

'Maaaaam,' came Edith's voice, bringing me back to my senses. 'Joe won't stand up.'

I didn't like the sound of this one bit; Joe was an old campaigner, we'd had him since Raven was little and he was no spring chicken when we got him. In the other stable, I found Joe alone. Della, the untameable free spirit, had gone. Joe was sitting with his legs tucked neatly under him, his head up. His hazel eyes were weary, and the coarse whiskers on his chin and the grey hairs that flecked his face showed his advancing years. Violet knelt beside him in the straw, while Edith stood over him with a pensive expression on her face.

'He won't get up,' she repeated.

Little Joe's ears flickered. He stared dead ahead but was listening hard.

'Food. We'll get him a bite to eat,' I said, making my way towards the proven store. 'See if that'll tempt him.'

There was a bag of mixed cereals and a drum of molasses in there, all edibles that Joe usually found irresistible. Food had always been his downfall, and, over the years, there had been many occasions during which he'd over indulged himself, though none so far had caused him to suffer any ill effects.

The prognosis for any animal that cannot, or will not, stand is not good. The greater the length of time spent as a 'downer' the more the chance of recovery diminishes; the patient would soon become weak, their stomach and digestive system begins to fail, and lying in soiled, pee-soaked straw was clearly not beneficial, so it was imperative that we got him to his feet.

'C'mon, Joe.' I shook the blue feed scoop that was now brimming with cereal flakes and the gloopy treacle. His ears quickly snapped back into forward position, and he craned his neck as I bent forward to give him a waft of the sweet-smelling muesli mixture to whet his appetite. I let him have a little taste, and a few flakes fell to the floor as he took a mouthful. A white cockerel that Miles called Brian, in honour of the famous scientist Professor Brian Cox, appeared out of nowhere and began scratching around amongst the straw bedding. Nobody, other than Miles, liked Brian as he was a habitual people-pecker. Miles had hand-reared him from an egg and now Brian ruled the roost, or so the feathered bird-brain thought.

I handed Violet the feed scoop and told her to hold it just out of reach of Joe whilst Edith and I pushed from behind. Even though he was only a Shetland pony, he weighed an absolute ton and other than him letting out a tired sigh from one end and a rumbustious fart from the other, there was little progress made.

'Shall we get Dad?' asked Edith.

'Christ no,' I said unequivocally. 'It's a no; Dad dislikes Little Joe very much.'

Then I ran through the list of reasons why.

'Joe scratches his bottom on drystone walls and then they fall down,' I said. 'Joe trod on Dad's glasses when he was tonsing tups in the barn. Joe chewed the wooden top-rail bar in the sheep pens. Joe kicked Dad's best tup when it tried to make a romantic gesture towards him in the field.'

I smiled at the thought of the ridiculously over-amorous tup that had attempted to mount Little Joe – and not in the conventional way that one should mount a pony. The tup liked Joe very much, but it was not a sentiment shared by our pony and we saw another side to him that day as, with a wild look in his eye, he bared his teeth, flattened his ears right down and repeatedly kicked out backwards with both back hooves. Joe was not shod and there was no lasting damage to the tup other than wounded pride.

That was when I remembered that the stock tups were in the loose box next door. I wondered whether a visit from one of them might serve as an incentive to get Joe up and about.

It worked a treat. No sooner had I manhandled the tup through the stable door than all hell broke loose. The tup let out a deep bellow of sexual interest, Brian squawked and flew up onto the hay manger above, and Edith and Violet recoiled to the back of the stable and watched the commotion unfold as Little Joe didn't just stand up, he leapt up as though he'd been given an electric shock. The tup ran a few circuits of the stable in a state of excitement, his nose in the air and his top lip curled back, while Little Joe spun like a top. I cornered the tup and frogmarched him to the loose box, then went back to the stable to assess Joe.

'Aw that brightened him up, Mam,' said Violet.

That was an understatement, as Joe was on high alert now. He gave a snort of indignation and looked furtively around, wary that the tup might make a reappearance.

This incident was the start of a new regime regarding Little Joe, as it brought to our attention the sad fact that this

233

lovable old rogue's health was now failing. A couple of ill-fitting rugs that swathed his low-slung frame were found in the blanket box in the woodshed loft. It took a few minor alterations, mostly requiring copious lengths of baler twine, to make them usable. As soon as I could get to the agricultural suppliers, I bought him some special veteran mixture, which was easy to chew and digest but not so easy on the pocket. However, it was worth the money as he consumed it very enthusiastically. We fed him apples, carrots and a hot bran gruel recipe that we found in an old horseman's book and, with this, Joe's health and strength seemed to improve.

Every day after the children returned from school, I'd see a small black figure cocooned in multiple rugs trot past the front door, closely followed by Sidney, who was running to keep up and laudably still holding on to the end of the lead rope despite nearly being pulled over. He would take Joe each night to be watered in the beck down by the bridge. There, with the water lapping at their heels, they'd stand together. Sometimes I could see Sidney chattering away; if nothing else, Joe still had a willing ear to listen to secrets.

Finally, the snow receded, but not for a good many weeks. The drifts of over twenty feet in depth that had accumulated in sunless valley bottoms gradually shrank and revealed the untold damage they had done. Lengths of drystone wall were razed, and water gates were washed away during the resulting floods. The snow storms had cost us dearly, left us with an enormous amount of clearing up to do, some epic feed bills and nearly killed us through sheer cold and fatigue, but we'd ridden out the storm. The animals were all safe and accounted for, we could move on.

We were at this time filming for television, talking about the impact of the storm, and had already explained that we were used to dealing with snow and bad weather, and that due to the accurate forecasting and our diligence all the flock were safe and well. I had sat on the quad bike surveying the scene as I waited for the camera to be set up again and for the soundman to get into position. Accompanied by Kate, the sheepdog, I'd just taken the Ravenseat flock of sheep back to their heaf at the moor, while Clive had been up to the allotment with his dog, Bill, to see whether the gate was still blocked with snow. Meltwaters had filled the beck, which flowed quickly and purposefully downstream. Ice still remained on the bankside and occasionally a lump would break free and be carried off over the ford and falls and away into oblivion. My eye was drawn by a chunk of floating wet ice in the middle of the beck which remained peculiarly still, never stirring as the current of water flowed around it. Kate's attention had also been piqued. I got off the bike to go and have a look, but wished that I hadn't.

'Rolling,' said the cameraman, panning around just in time to capture the look of dismay on my face as I saw that the mini iceberg was actually the ice-encrusted fleece of a dead sheep. I could just make out her bloated form under the peaty water. At this point, Clive turned up too.

'Rollin' down the river,' piped up Clive as I scowled at him. It wasn't funny. I had just talked in great depth about how we had not lost any sheep in the storm and now here I was, on camera, fishing one out of the river. It wasn't even as though I could claim that it wasn't one of ours because it was clear, owing to its location, that there was no way that it could possibly be anyone else's. I could only think that when Miles and I had brought the sheep down from the moor, one of the yows had been left behind. She must have sheltered beside the river and been overblown, then come to rest under the water and been covered in ice and snow. It was all a little humiliating, but it did reflect the true cost of the storm and brought the realities of it right into the living rooms of the viewers in high-definition technicolour.

It was a cause for celebration when, finally, spring arrived, and the sunshine brought a spark of life back to the bare pastures. Grass (or Doctor Green, as it is sometimes called) was the long-awaited cure-all and it felt like nothing short of miraculous when turn-out day came, and Joe went out to grass for the first time after what had seemed like the longest winter. We were jubilant, or should I say that the children and I were.

'Nay, I thought t'owd beggar was dun' for,' Clive said one day as he watched Clemmie sharing an apple with him.

'Just needed a bit of TLC,' I said. 'I'll do same for thee when it's time to put you out to pasture.'

It was a funny thing but all throughout that protracted winter it was the thought of Little Joe having one last summer during which he could feel the sun on his back, the grass beneath his hooves and truly enjoy the freedom of retirement, that kept us all positive. That was the goal. I had warned the children that it was probably going to be his last summer and that if he lost condition before the onset of the next winter then they must be willing to accept his fate. To put him to sleep would be the kindest thing to do, the hardest thing also but it would be the right option. Better than watching his health fail and seeing him suffering and in pain.

Now as I write, in November 2018, the first frosts of winter have arrived, and the trees have shed their leaves, a cool wind blows and the horses' glossy, sleek summer coats have been replaced with the thick shaggy variety. Little Joe is looking well, once again wrapped up against the elements and consuming feed with vigour, and we will endeavour to nurse him through another winter as this spirited little pony is clearly not ready to give up the ghost just yet.

There are only a few folk left who can remember when working horses were used on farms. Our horses are for pleasure, rather than work, but, in the same way the children like to help, there does seem to be a deep-seated desire within them to contribute, to have a job. Josie and Princess are not particularly well schooled, would fail in any dressage test, but they are a useful mode of transport, being surefooted and quick learners. I think they take great enjoyment in being part of the team. Hopefully, we will get Della broken in, so we can ride her, although she may be a little more of a handful owing to her shy and nervy disposition, though gentle coaxing will undoubtedly win over her insecurities.

For the horses, summertime is their holiday, when they can graze away at the moors and be wild and free. Wintertime is about routine, being stabled and ridden – but in the most casual manner possible – around the sheep or perhaps to the moor, nothing too strenuous. The Firs has old stables that stand across from the house, opening out onto a little stone-flagged yard, and we took Josie and Princess down there for a few weeks, using them to explore the surrounding fells and moors.

Our old friend Johnny was a good hand with horses. Although he had retired from farming, he had never parted with any of his tack and was happy to lend us items from his substantial collection, as well as being on hand with advice, useful or otherwise.

He was well into his eighties, a diminutive, wiry man but a noticeable figure in that he was a snappy dresser. As well as his flat hat, houndstooth blazer and brown dealer boots, he had a penchant for yellow cord trousers with extremely high waistbands, coupled with braces, which seemed to accentuate his pronounced bowed legs.

'Used to ride out for Captain Crump,' he'd sometimes say. A silver-framed picture of him on the sideboard showed him in his younger years on a thoroughbred at a point-to-point cross-country race in a pair of voluminous elephant-ear breeches and gaiters. Now troubled with arthritis and rheumatism, his fingers were twisted and bent inwards on themselves almost as though he was still holding reins, but what a wicked glint he had in his eye and he wore a half-smile that would effortlessly blossom into a beaming grin. I did sometimes wonder whether the tomfoolery and outward smiles hid a deeper unspoken sadness, for I knew that back in the day there had been a child stillborn. It was a subject we never broached with either him or his wife, Dora, it was just one of those unsaid things that everybody knew about but nobody mentioned. It appeared that they filled this missing part in their lives with animals, and had farmed sheep and cattle and bred

Dales ponies but now, as old age set in, the farm and land was all let. They lived in a comfortable bungalow nearby, with a paddock, enough room for a couple of ponies, and the fancy hens, pigeons and cats that they loved.

Dora wore a headscarf at all times, knotted under her chin; I don't think I'd ever seen her without it. The auburn permed hair that it covered never seemed to move. I'd considered that it might just be a wig, but if it was then she wore it well. She'd stand on the doorstep, hands shoved down into the front pocket of one of her vast collection of gaily patterned aprons, with colourful plastic flowers in hanging baskets on either side of her. In her garden, random objects, like wheelbarrows and plastic feed buckets, were all brimming with plants and surrounded by chicken wire to prevent the chickens that wandered freely around from spoiling the displays. The household seemed to accumulate cats. She even offered her services as a cat sitter and would take care of other folks' feline friends whilst they were on holiday or in hospital. Clive and I had dropped in one afternoon with the intention of borrowing a mouthing bit for Princess, a nice soft bit with metal keys for her to play with using her tongue. When we arrived, there was a very well-heeled lady giving Dora and Johnny specific instructions on how to look after her beloved cat, which I assumed was the one in the wicker basket on the floor.

'Here's a week's worth of food,' said the cat owner. 'A tin a day, if you will.'

She placed a plastic bag on the table, going on to detail the moggy's daily habits before she finally departed, dabbing her eyes with a carefully folded handkerchief. She'd hardly been gone two minutes when Johnny decided to take a look at what was in the bag.

'Salmon,' he said, picking up one of the tins and admiring the label. 'Weeel, that's us sorted for the week, Dora. Kitekat for you,' he added, tapping the wicker basket, from which a mewing sound ensued.

Dora shook her head, but she, too, had a twinkling smile full of devilment.

They were honest folk, salt of the earth you might say, but Johnny – having been involved with horses and trading all his life – was exceptionally sharp-minded and always on the lookout for a deal. Every year he went to the annual Cowper Day horse sale at Kirkby Stephen. He could always be found in the auction mart watching the trade and wooing potential buyers. I'd seen him reel in the customers firsthand.

'I'm wanting a pony for my daughter to ride,' said a besuited gentleman who looked just a little incongruous among the gypsies and horse dealers, teenage lads with mullets and girls with fake tans and talon-like nails.

'Aye, I think that I's got just the pony for yer at home,' Johnny'd said, nodding his head. 'It's 13.2 or thereabouts, gelding, quiet in all yokes, ride n' drive, bombproof, lovely lal' 'oss.'

'And how much would you be wanting for it?' asked the prospective buyer.

'I'd tak five hundred for it,' said Johnny confidently.

'Oh, well,' said the gentleman, pausing for a moment. 'I was looking for something maybe a little bit better.'

'Weeeell,' said Johnny, thinking quickly, 'I do hev another, but I wasn't really wantin' to part wi' it.' He had his buyer under his spell, the trap was set and the man was going to take the bait. 'Proper genuine sort, but it'd 'ave to be fifteen 'undred.'

I had a sneaking suspicion that these two horses were one and the same.

We called on them once when Violet was a baby. I'd just unfastened her seatbelt and lifted her out of her car seat when Johnny came over and reached out to hold her. He walked towards the kitchen window, where Dora was doing the dishes.

'Look what I getten,' he'd shouted, holding Violet up to the window. 'I've swapped it for a bantam.'

Dora laughed, and if she wasn't really laughing inside then she did a jolly good job of hiding it. The older generation maybe had a different way of dealing with such painful subjects.

We did do a swap that day, but it was hatching eggs that changed hands rather than a baby. Johnny wanted Welsummer eggs to hatch under a broody hen; I had some of them and in exchange I got a mixed batch of hatching eggs from his fancy bantams to put in our incubator. In fact, my small incubator was such a notoriously awkward thing to get exactly right that I didn't hold any real hope of success.

I was wrong, they all hatched: two silver-spangled Hamburgs, a blue Wyandotte, two silver Sebrights. We reared them all. Miles took the silver-spangled Hamburg to Muker Show, which at the time had a poultry section. Although Miles had the smallest chicken at the show, he won each of his classes: best local chicken, best small-breed chicken and overall best chicken in show. Rather like Johnny, what that little chicken lacked in stature he made up for in personality and attitude.

8

The Adventures of Chalky

Late March can be filled with seemingly endless grey days, and in 2018 the month was, as usual, wet underfoot with little prospect of anything better. Our workload also increases at this time, as the sheep need to be tended more frequently owing to the imminent arrival of the lambs. In an effort to maximize efficiency and get around all the heafs quicker, we had invested in another quad bike. Now Clive could go in one direction to his flock on Ravenseat moor, whilst I could go the other way, to Black Howe and Birkdale Common, to mine. Every morning we would load up with hay and feed for the sheep, and in my case three children, Annas, Clemmie and Nancy. They would hunker down in the trailer between the hay bales and bags of cake, peeping out from beneath their woolly hats. Neither Clive nor I would take a sheepdog with us, because at this late stage in the sheep's pregnancy the last thing they needed was any kind of upset or stress.

The hard work of lambing time was just around the corner, and it would be then that the sheepdogs would be invaluable, so for the moment they could enjoy some well-earned time off. However, Pippen and Chalky, as terriers of leisure rather than working dogs, chose to snooze whenever the fancy took them. Often, they could be found laid out on the hearth beside the fire in the afternoons and early evening. At nighttime they would come to life, baying at the door at last light, ready to head out into the dusk to patrol the farmyard. In the morning they would return to the farmhouse door, sometimes scarred with snout wounds that told of nighttime battles with rats. After a brief interlude patrolling the kitchen in search of discarded breakfast food, they'd loiter around the yard awaiting the departure of the first available quad bike. Quite why they liked to gallop all the way to the moor and then almost immediately turn around and gallop all the way back was a mystery to me, but every morning it happened. Neither dog was fast enough to keep up with the quad bike and being on a short leg meant that they often needed to take detours. Becks and gutters that we crossed with ease would have to be carefully negotiated by the dogs via stepping stones. Sometimes I would reckon to have shaken them off, but without fail they'd show up, often wet and panting from sheer exertion. The sheep weren't too bothered about the terriers, they'd be far too interested in the impending meal of hay and sheep nuts to give more than a casual glance in their direction or the occasional look of disdain if one ventured too close. Clive would get annoyed with the terrier twosome and growl at them under his breath, telling them to 'Git away yam'.

As the terriers were free spirits, nobody ever really took

notice of their whereabouts. They were just always in the background, checking in and out as and when they wished. Pippen, being the older of the two, was more set in her ways and less open to persuasion than her counterpart Chalky. Pippen's days of being led astray were pretty much over, her final foray into the unknown being a trip over the moors to a far-flung spot where she became entangled in a long-forgotten snare. This journey had almost been a one-way ticket as it was two days later when we found her, head bloated and eyes bulging, the wire noose slowly and painfully strangling the life out of her. She recovered, but was left with a bald ring around her neck where no fur grew. It spelled the end of any great excursions and now she never strayed too far from her own patch, enjoying her home comforts far too much to consider taking off.

Chalky, on the other hand, was as fickle as could be and would think nothing about heading out into foreign climes without so much as a backward glance. Being a sociable creature she enjoyed the attention that the passing walkers lavished upon her – and any food they might have – and would follow them on to wherever they were heading. For six months of the year, when Ravenseat was bypassed by walkers and visitors and reverted to being a rural backwater, there was no temptation for Chalky to leave us. Indeed, there were rich pickings soon to be enjoyed around the farm at lambing time and I am sure that Chalky was well aware of this. The placenta expelled by the sheep after the birth of the lambs was one of her favourite snacks. It was entirely natural that a dog should enjoy eating these as they were wholesome, good and packed full of nutrients, but it still turned the stomach to see her exaggerated chewing on the bloody membranes.

So, it came as a bit of a surprise when we realized that Chalky was missing.

'Has she not bin wi' you to t'sheep?' I asked Clive.

'Nay, brown 'un 'as,' he said. 'Mebbe, I dunno, I cannot remember.'

I'm ashamed to say that neither of us could recall when we had last seen Chalky. With no confirmed sightings from the children in two days, and no clue where to even start, the search began. We checked around the farmyard, under the feeders, in the hay lofts and every outbuilding, but there was no sign of Chalky. We looked in ridiculous places: right under our noses, in the hall robe cupboard, the odd-sock bag where occasionally Chalky had curled up and slept, behind the sofa and in the place that nobody *ever* went, under teenager Raven's bed, where even angels fear to tread. As the search around the farm proved fruitless, we cast our net further afield, this time visiting our most far-flung barns and stone-built shelters, always hopeful that from behind a rotting wooden door we might hear a faint whimpering.

We tried the local hostelries – Tan Hill Inn, The Farmer's Arms at Muker, The King's Head at Gunnerside – where she'd turned up in the past, but nobody had seen her. We looked at other favourite haunts, the local campsites, which were still deserted owing to the cold spring, and we asked a few gamekeepers who tramped the moors and desolate places but, again, nothing.

Frustratingly, Pippen seemed unconcerned for the welfare of her partner in crime, she just stretched herself out across the cushion that they had previously shared together, luxuriating in the newfound roominess of the gap beneath the long settle.

As time went by, I'd take a slightly different route every day when travelling out to the sheep, convinced that Chalky might be right under my nose and that I had just never seen her. It was a ridiculous notion, but even though I had resigned myself by now to looking for a body, I still needed that closure. It was vexing to not know what had happened.

Understandably, the younger children were upset, but the older ones displayed a more stoical acceptance of the situation.

'She 'ad a perfect dog life,' was Reuben's take on Chalky's demise.

'We should get another dog,' said Miles. 'Pippen will be lonely.'

'Pippen ain't lonely,' retorted Raven. 'She's lovin' it, more space, more dropped food to hoover up, it's a win-win.'

'Look, it's been two weeks now an' there's bin no word 'bout Chalky,' said Clive. 'It's time we drew a line under it and accepted that Chalky has gone forever.'

'Dead,' said Annas, who had an obsession with death at that point.

'Yes, dead,' said Clive.

I nodded. 'I'm gonna write the Twitter obituary,' I said. 'There's no hope now.'

Commiserations came in thick and fast, phone calls from friends and messages from people far and wide who knew the pain of losing a pet, but life goes on and that was that. In the run up to lambing time there was a huge amount of walling to do. Making the fields stockproof before we gathered the sheep down from the moor was a matter of urgency and Clive and myself were desperately struggling with the huge workload.

'I'm gonna ask Tuppence if he'll gi' us a day or two wallin',' Clive announced over breakfast one day. 'We need to get a straight edge afore lambin' an' we're runnin' out of time.'

I was in total agreement. Tuppence turned up the next morning, striding into the yard in his standard-issue long black railway coat and hobnailed boots laced to the turned-up toes. A long-standing friend of Clive's, he was a countryman through and through, a master craftsman in the art of walling and hedging and extremely knowledgeable about birds and wildlife. He was strongly built, though not heavy-set. Pale-skinned and thin-lipped, his deep-set piercing blue eyes darted this way and that, taking all in. He stood tall, though slightly stooped, leaning forward, like someone bracing himself against the wind. One hand cupped the top of a knotty stick that, like himself, had a slight curvature, and wrapped around his other hand was a length of twine, on the end of which was a broad-chested wiry-coated terrier.

Tuppence was in his eighties, but still as strong as a man half his age. He attributed his good health to fresh air, being out in the elements and walking. Not just gentle ambles in the countryside, but traversing hills and slopes to get to the most awkward of places and then spending the day lifting rocks and boulders and fitting them together so beautifully that you could hardly tell that there'd been a repair in the wall at all.

'Whaaaaat,' was his standard greeting, to which he got the standard reply from Clive.

'Aye, owsta keepin'? Yer lookin' lish like.'

Farming can be relentless . . .
but we wouldn't change it for anything.

(*above*) Jack o t'Firs watched over our renovations from above the fireplace.

(*right*) A medieval instrument of torture or a cheese press? I searched high and low for this antique.

(*below*) What home is complete without an organ?

The kitchen is the heart of any farmhouse.

The infamous half tester bed at The Firs.

(*above left*) The fodder-beet lorry tipped on its side after hitting ice.
Moving the beets by hand was not fun.

(*above right*) When the Beast from the East hit, we had to take
feed to the sheep by hand, wading knee-deep in snow.

(*below*) Kate, her face encrusted with snow. A shepherd is nothing without their dog.

(*above*) With not a blade of grass in sight, the sheep were totally reliant on us to bring them food.

(*right*) Even in the harshest weather, there is still beauty to be found.

(*above left*) Chalky on her way back home after disappearing for weeks.

(*above right*) Bill, an old fella now, watching over Nancy.

(*right*) Summer 2018 was all about wild swimming. Clive leads the younger children to the river.

I took the older children on an adventure,
swimming in the river at Boggle Hole.

(*above*) We spent days searching for a missing calf after her mother was found dead. Here the children head off onto the moor tops.

(*below left*) We found the calf, Joy, the very night we were about to stop looking. We brought her inside to gently warm up by the fire.

(*below right*) One year later, Raven took this picture of Joy with the other cows at The Firs.

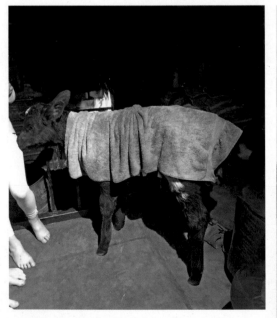

'I's reet, what aboot thoo's gurt dog?'

'What, Bill? Nae t'hell, ee's gan' reather porped.'

'Better it than thee.'

A conversation that, to the outsider, would have needed a little translation.

Tuppence tied his dog to the gate and came into the kitchen to get instructions as to where the gaps that needed mending were. I made him a cup of coffee.

'What aboot thi' bait?' asked Clive. 'You've gotta bit of a way to t'first spot.'

Tuppence lifted the flap of the pocket of his coat and proceeded to rootle about inside. Out came a plastic bag in which was a meagre slice of dry white bread and what looked like a bit of cheese.

'You on rations or summat?' said Clive, looking unimpressed. ''Ere, 'ave a scone wi' yer coffee.'

I dutifully handed over a scone out of the tin on the side and they talked for a little bit on the usual subjects: weather, trade at the auction and who was 'in a bad way'. Tuppence generated crumbs which dropped on the floor.

'Where's thi' dog at?' he asked, staring at the crumbs.

'Chalky's a-wantin', dunno what 'appened, been gone for two weeks now,' I said flatly.

He gave a protracted, 'Ayyyye.'

'We've all been gae dowly, I'll tell ya,' I said.

'Bloody thing,' said Clive crossly, ''t only 'ad itself to please.'

There followed a long silence.

''Ad it got brown spots on its lugs?' Tuppence asked.

I said that yes, she had little flecks of brown on the tips of her ears but that they were not really that visible.

'Why?' I asked.

'I think I've seen thi dog,' he said ever so slowly.

'*Where?*'

My mouth had dropped and even Clive's face had taken on a look of complete astonishment.

'Face beeak,' Tuppence announced in the thickest Dales dialect imaginable.

'*Facebook!* You're on Facebook . . . and more to the point, mi dog is on Facebook? I's not even on Facebook.'

'Aye.'

I could not believe firstly that Tuppence had an online presence and secondly that he'd seen Chalky on there.

He explained that a couple of weeks ago he'd seen a picture of a small white terrier with a wiry coat and speckled brown ears that had turned up at a pub on the outskirts of Kirkby Stephen.

This already sounded promising, with Chalky's history of frequenting boozers. The dog had no collar and had been taken to the local vet to be scanned for a microchip, but it failed to read. She had then been taken in by a local farmer and subsequently rehomed and was currently enjoying the high life as a pet dog. She had been shampooed and groomed to within an inch of her life and was now living under the assumed name of Twiglet!

Chalky was alive and well only some fifteen miles away, and if it had not been for Tuppence then we would never have known. I rang the veterinary surgery, and they confirmed what had happened and arranged for Chalky to be returned after being microchipped again. We can only guess that Chalky had set off on the Coast to Coast footpath but instead of heading down Swaledale she went in the opposite direction to Cumbria. We know that she had a collar with her name tag and phone number but must assume that at some point it became detached.

It was a cause for great celebration at Ravenseat when the prodigal terrier returned. Not with her tail between her legs, though – the moment she returned she scampered back off into the barn, tail aloft and teeth bared, ready to continue the war against rats. Neither did Pippen seem that elated; after the obligatory sniffing of behinds it was back to business as usual, sleeping, eating and fighting.

I had always liked Tuppence but after this he went right up in my estimation. It broadened the topics of conversation too. Where once we would talk only of the weather and whose sheep were looking in good fettle, we could now discuss global affairs and politics and it extended my reach on the gossip front, as I gleaned information from his newsfeed. Poor Tuppence didn't get long basking in the glory of his online detective skills before the tables were turned. It was his third consecutive day of gapping at Ravenseat when he came back into the farmyard dogless.

'What 'appened?' I said.

'Nae, Rusty's gawn, tekken off,' he said breathlessly, his normally pale cheeks red with exertion.

Tuppence, as a rule, never let Rusty off his lead because the dog had a blinding obsession with rabbits and, given the chance, would pursue them relentlessly until he either caught one or it went down a burrow. Then things would get worse as Rusty would never admit defeat and would go to ground. A spade would then be required to dig him out and this was only possible if you knew which tunnel he'd gone down.

'Where? An' which way did 'e gan?' I asked.

'Over yonder,' he said, waving his stick towards the Close Hills pastures. 'I was eatin' mi bait, Rusty was off 'is string an' then 'e put up a rabbit.'

He delved around in his pocket and eventually his hand emerged clutching an old mobile phone.

'It won't work,' I said, 'there's no reception around 'ere.'

'Naw, I want thoo to ring mi missus on t'ouse phone an' tell 'er I's gonna be leeate back.'

I did as he asked, left a message on the answerphone and then set off to the allotment, the best vantage point to watch for sheep moving in the distance, which was usually a sure sign that there was a dog on the loose. Tuppence returned to his gap, this time with Clive on the quad bike.

Rusty had disappeared without trace and, in true Chalky-style, there was no sign of him anywhere. The children returned from school and decided that they, too, would join the search. They knew all the likely places, after all this was the second time in as many weeks that we'd been out hunting for a lost terrier. Finally, when dusk came, Tuppence bade goodbye. He was not in the best frame of mind, muttering under his breath, cussing his dog and looking troubled. He was now without the long black coat which, upon Clive's advice, had been rolled up and laid just inside the door of the barn nearest to where Rusty had vanished.

'He just might land back to t'place he last saw his master and settle for t'coat being the nearest comforting presence.'

I'd heard this theory before, spoken as a kind of folklore

amongst dog owners who'd recount stories of sheepdogs being sold on after their owner's deaths only to refuse to run or be in any way obedient. The solution was always to take an item of clothing that previously belonged to the deceased owner to give the dog the familiarity that it needed in order to relax and understand.

'Maureen's gonna kill mi,' Tuppence said to Clive. 'She dotes on that lal' dog.'

'More than thee?' said Clive.

Tuppence didn't answer.

'I'll be back first thing in t'morning,' he said heavily as he got into his car.

Sure enough, the next morning the barking of the sheepdogs in the farmyard told us that someone was around. I lay in bed, not willing to move, while Clive stirred and mumbled. Pebbles hit the glass of the bedroom window. Clive staggered out of bed and went to look who was outside.

'Whaaaat,' he shouted, mimicking Tuppence's usual greeting.

'Is ta gonna lig in bed til' sun burns yer eyes out?' replied the voice below.

'Yer just up an' about early 'cos you've been gettin' some earache,' retorted Clive.

Rusty had not returned and we were now at a loss as to where to look. I had to go to Hawes to get some bags of hen feed and lambing-time supplies, so loaded the little ones into the pickup promising them we'd call at the sweet shop whilst out. The morale boost would be needed if the day was going to be spent searching for the lost dog. We set off in no particular hurry, the children looking out of the windows and pointing at various things that caught their eye. It was the usual stuff.

'Sheeeeeep,' said Clemmie.

'Yep, soon be lambs too,' I said, staring fixedly ahead.

I tried to sound upbeat and enthusiastic when actually the mere thought of lambing time tired me; it was taking us the entire day just to get all the animals fed and seen to as it was, so goodness only knows how it was going to be when the lambs began to arrive. The relentlessness of the hard winter, preceded by a dire summer, had taken its toll and we all, the animals included, needed some sunshine.

'Peewit,' squealed Annas, delighted to spot a lapwing nervously strutting at the roadside before taking to the air in fright at the roar of the engine as we passed by.

It can be easy to overlook the sights that pass you by every day, so to see the children's wonder was heartening. 'Nebbin' was what I was doing as I crossed the Hoggarths Bridge looking up towards the Keldside Allotments, where we had a handful of shearlings grazing.

'Dog,' shouted Annas.

It didn't register.

'Daaawg?' piped up Clemmie, more quizzically.

I quickly swung round to look at the road ahead and saw Rusty sniffing about beneath the budding hedgerow by the verge. I slowed down and pulled to a stop at the side of the road. I wasn't sure whether Rusty was biddable, so needed to tread carefully to avoid spooking him. I gestured to the children to be quiet as, by now, I could see their faces pressed up against the back window as they waited to see whether the ambush was going to be successful.

I need not have worried, as Rusty came right to me, his short, stumped tail wagging furiously back and forth. He looked up at me expectantly with shining dark eyes, his muzzle wet from rootling around in the undergrowth. I picked him up and carried him to the pickup, noticing he was quite a weight for a small dog, solid as they say. The children were delighted with the new passenger, who chose not to sit in the footwell but to balance precariously with his front paws on the dashboard and hind legs on the passenger's seat. He stared intently out of the windscreen as I turned around and went back to Ravenseat.

This time it was Tuppence's turn to be grateful. 'Yer bogger yer,' he growled as the pair were reunited. 'What trouble I've bin in 'cos of thee.' And, smiling, he scratched the back of Rusty's head and attached the string to his collar.

In our line of work, the dog is not only our best friend but also a colleague. The saying 'what is a shepherd without a dog' is certainly as relevant as ever in the hills. Quad bikes are often quoted as being the death knell of the sheepdog, but in the most inaccessible and inhospitable areas the sheepdog rules. They are a tool of the trade and an invaluable asset to the hill shepherd; never should there be a time when you do not have a four-legged companion with boundless enthusiasm to accompany you throughout your daily tasks. Inevitably, the passage of time dictates that throughout the course of a lifetime you will be accompanied by a succession of dogs. Some you will recall fondly, others are remembered for their failings and mannerisms.

Bill, Clive's stalwart dog, is now firmly in the realms of

the elderly. He has served his time but shows no signs of wishing to retire from the business of keeping order amongst the sheep. Loving and playful when in my company, belligerent and truculent when with Clive, he is a clever dog with a split personality. In his younger days, his athleticism was to be marvelled at. There was no wall he could not scale, no river he would not cross, no obstacle that could be put in the way of Bill and the rounding up of sheep. Bill has learnt every gather, how and where the sheep run, the places where the sheep can get away. My dog Kate is younger, but she, too, now knows the lie of the land and, although her and Bill are as different as chalk and cheese, they do complement each other. Her energy makes up for his lethargy. Bill's mind is as sharp as a needle, but now his body is beginning to fail him; Kate is in her prime but lacks the same intellectual qualities. Together they are the dream team.

On a big gather, when we bring the flocks down from the moor, we'll ask Clive's friend Alec to come along with his sheepdogs and help us get them back home. It is tiring work so extra dog power will never go amiss. This spring, talk got around to our need for up-and-coming young sheepdogs ready to take over the mantle of top dog. It is not something that anyone likes to dwell upon but, nevertheless, it is a fact that the day will come when Bill cannot cope with the workload.

'Yer need to be gitten thi' sel a young dog boyo,' were Alec's words of wisdom to Clive.

'Yer do an' all,' I chipped in.

'I'll take Kate if I 'ave to,' Clive replied. This was met with the sternest of looks from me.

'She's a one-woman dog an' I don't think that she'd run for yer,' I said. It is a most difficult thing to persuade a sheepdog to run for someone else, especially when it's older and already set in its ways

'Worra bout yon' woolly dog?'

Fan, our young bearded collie, seemed as keen as mustard, but there was little work in her – she would run for a short while then become uninterested. She was destined to be a part-time sheepdog and found her way, via Alec, to a small-holder in Wales who needed a relaxed dog to move a few sheep around his pastures and be on hand to chase the sheep through the pens.

An opportunity to get a new dog arose when Ben, one of our neighbouring farmers, asked if it would be possible to line his best working bitch with Bill. 'You'll get pick of the litter' was the agreement. The liaison resulted in two puppies being born, both dogs. Ours, we christened Roy, in the vague hope that it might follow in the footsteps of one of Clive's best-ever sheepdogs. So good was the original Roy that when Clive and I began dating he made it abso-lutely clear that if Roy did not 'tek to mi' then our budding relationship could not progress any further. I loved the fact that his dog's happiness mattered so much, but I admit that I was also relieved when Roy gave me the seal of approval by looking up at me from beneath his bushy tan eyebrows and lifting his head towards me in an invitation to stroke it.

Young Roy is a classic border collie with a flowing black-and-white coat. He has a habit of cocking his head to one side when listening to you and has inexhaustible energy that he chooses to expend rounding up sheep, so the initial signs are good.

Roy is the children's friend, will go on walks, and has even tolerated a shampoo in the shower, but Clive's big bugbear is that he has a propensity for jumping up. Like a coiled spring, he bounces with great big paws that reach almost to your chest. It is frustrating that he cannot be dissuaded from doing this, but it does sum up his whole demeanour, which is one of bubbly, joyous happiness.

I do understand why Clive prefers the more aloof stance taken by most working sheepdogs. They are a breed apart from a pet dog and their entire focus is on sheep; they seek to please, of course, but their pleasure is taken from the thrill of the chase. You'll never see a shepherd showering his loyal sidekick with animated belly rubs or being overly affectionate, it's more about a subtle word or casual touch on the head, a gesture that is a show of respect.

'Dun't black 'im,' I'd say to Clive when Roy would launch himself at me, paws invariably caked in mud, his great face smiling, tongue lolling and tail wagging with unadulterated pleasure.

'That dog is an idiot,' he'd say, shaking his head, and it was true, but I liked him.

Since Roy was no more likely to change his ways than Clive, we needed a new dog, a younger version of Bill who could gradually take his place doing the harder work. Bill could still be used where intricacy and precision were needed. Perhaps in a roundabout way Bill would appreciate having to work a little less hard and be able to take more enjoyment in doing his job, maybe not turning out on the wettest, coldest days and leaving his younger counterpart to learn the ropes and develop a relationship with Clive.

Clive found himself a dog called Joe, who was big, ugly and unbelievably keen. His capacity for work was second to none. He'd have been brilliant if he had a clue what he was doing. In essence, he was insane, he just wanted to run and run and run. Even in the kennel he would run round and round until his paws became sore. He needed to be distracted, he couldn't control himself.

Clive had a brainwave: to bring him into the yard and put him on a chain, and that way his head would be occupied with watching everyday life unfolding before his very eyes. The chickens that wander around the farmyard gave him a very wide berth, they could spot a nutcase at twenty paces. The peacock also stayed well clear, though the children would go across and give him biscuits and leftovers and never once did he show them any malice.

It was sheep that drove him wild but, although the general idea was that he should chase them, Joe did not know when to stop. To take him to the fields amongst the sheep was stressful. We would even run him up and down the road, with him following the quad bike, just to try and cool him off, but nothing seemed to take the edge off him. We had to buy him a muzzle when he committed the ultimate crime of sinking his teeth into the shank of a sheep that stood its ground.

We bemoaned the fact that we had one soft dog that bounced and rolled over and behaved like the Andrex puppy and another that was only one step off the Hound of the Baskervilles.

The upside of these dog issues was that Bill had developed an extra spring in his step; a new-found energy coursed through his veins, fuelled by loathing. Rather than welcoming the arrival of a new dog on the kennel block it had unleashed an intense hatred within him. Roy, his own flesh and blood, the fruit of his loins, he was all right with. In fact, no man nor beast could dislike him, but Joe was an entirely different matter. To say there was no love lost would be an under-statement, though I had no idea how vehemently Bill despised Joe until I made the mistake of taking them all on a little walk up the sheep-less moor bottom. Edith and I opened the mesh doors of the kennels and all four dogs galloped off towards the beck. We caught up with the dogs and sat down on the bankside where Roy was bouncing enthusiastically around Kate, doing his utmost to impress her whilst she flirted back and then occasionally snapped at him if he got too close. Bill lapped water from the beck and quietly watched the shenanigans, seemingly unimpressed. Joe had already done numerous circuits around us and the other dogs, in search of anything to chase, but having failed to find suitable quarry, he now amused himself with the next best thing to a sheep, sheep droppings. He nosed through them with considerable enthusiasm while Bill looked on disconsolately, his ageing eyes unmoved and emotionless. Joe was oblivious to the fact that his every move was being scrutinized from afar, and Roy and Kate were too engrossed in each other to notice that a storm was brewing. By the time I realized that Bill's hackles were rising, it was too late.

'Let's go yam,' I said to Edith, thinking that I could diffuse the situation.

We stood up and turned for home, Roy and Kate galloping side by side back towards the kennels. Before I could even whistle for Bill and Joe, I heard the horrific sounds of a dog fight. Snarling, they locked jaws, blood pouring from both their faces, while Edith stood motionless, screaming. Bill had taken on a whole different persona and was now biting Joe like a dog possessed.

'Do summat!' screamed Edith as I stood transfixed. 'Bill's gonna get hurt.'

Adrenalin must have been coursing through Bill's veins because he was relentless in his attack; if there was going to be any winner in this battle then it was going to be the elderly statesman, Bill.

'Ga an' get a stick,' I shouted at Edith as I edged closer to the warring pair. 'Hurry up, please hurry up.'

As Edith hurtled off back to the kennels, I picked up a large topstone that had fallen from the wall of the sheep stell (a circular stone pen) beside the beck. I flung it as near to the pair as I dared, in the hope that it might distract Bill and persuade him to let go of Joe, who by now was upside down, prostrate on the shingle beside the water. Bill was standing over him, his tail on end, with his jaws clamped tightly around the loose skin at the back of Joe's neck. Joe's head was screwed around so he was looking upwards at his towering aggressor, yelping in pain.

Bill let go and flinched as the stone flew past and splashed into the water.

'That'll do, Bill,' I shouted. 'That'll do, yer worrying beggar.'

I could see Edith coming back with a shepherd's crook, accompanied by Reuben and Miles.

Bill was breathing heavily, his chest heaving, as he stood triumphantly over his fallen enemy. Joe stared fixedly back at Bill with wide eyes.

'Bill, that'll do,' I yelled.

He woke from his stupor, only to resume the attack. This time Bill dragging a flailing Joe by his shoulder towards the beck. Joe retaliated by sinking his teeth into Bill's ear. The splashing, growling, blood-and-fur flying was a terrifying spectacle and one that I never want to see again.

'Quick, stop 'em!' shouted Reuben, who snatched the crook from Edith and thrust it at me.

Edging forward to the beck, I reached out and slipped the hook end through Bill's collar, yanking him sharply towards me. For a few seconds, he kept his stranglehold on Joe, but then he reluctantly released his grip and instead angrily turned his attention to the crook, pulling against it with all his might. I prayed that the collar would not snap as I dragged Bill out of the water. His attention was still on Joe, who now stood, sodden and forlornly panting in the water, a look of dejection on his face.

The children were standing open-mouthed behind me. To see Bill, a dog they have known all his life, change like that was shocking.

'Bloody 'ell, that were nasty,' said Reuben.

'Nivver underestimate the damage a dog can do,' I said.

Reuben called Joe, who had had the wind knocked right out of his sails. I stood back, with Bill still anchored on the end of the crook.

Joe, being the younger dog, had withstood the attack better than it had first seemed, and only had a few superficial injuries, a tear in his ear and puncture wounds around the scruff of his neck. They had bled plenty but it had all looked a lot worse than it really was. There was nothing on him that required stitching, just the application of some antiseptic salve to ward off infection.

'Poor, poor lad,' said Edith to Joe, who had a hangdog expression to beat all others, as the children walked back with him to the kennels, where I could see Roy and Kate were waiting to be let back in.

'I'll land in a minute,' I shouted to them.

Bill's blood was still up, though the fact that Joe had departed had calmed him down considerably, to the point where I now dared lay my hands upon him. There had never been a time when I had any fear of Bill, not once had there been any sign of malevolence in him. I would say that I trusted him implicitly and I still do but to see him turn like that, even against another dog, was an insight as to what basic instincts can remain hidden from plain view. Although Bill was the victor of this dog fight, he had sustained a few injuries, losing a tooth and now walking with a pronounced limp. I know now that it was entirely my fault and that I should have recognized that Bill felt Joe represented a challenge to his position of authority, and that in order to remain top dog he needed to show Joe exactly who was the boss.

I vowed never to let the two out together ever again.

After the dog fight, we all went through a period of calm. Joe and Bill rested and recuperated, and I thanked my lucky stars that it hadn't ended with a fatality. When Joe did return to work it seemed that his confidence had taken a battering and that he was not as bold as he was prior to the fight.

Frankly, he was better for it. The muzzle came off and, although still keen enough on rounding up the sheep, he was now controllable. Clive began taking Joe out and to work. A trust was growing between the two and a bond began to develop.

Bill now only came out to work on his own or with Kate. As far as we were concerned, Bill, though old, was still the alpha male, the top dog who possessed the most knowledge about our farm and operations and would never fail us.

It was a beautiful warm May evening when Clive and I decided that we would go to the allotment to push the last of the yows and lambs through the moor gate. We had spent the morning marking them and tagging the lambs, then had taken them to the allotment to mother-up and graze before the final part of the journey to the open moors. I'd spent the afternoon serving teas to the walkers who passed through the farm on the Coast to Coast, whilst Clive had been mucking out one of the buildings in the farmyard. Now that all of the animals were out in the fields, it was a good time to get the buildings clean and tidy and do a little maintenance around the place before the next big tasks of clipping and haytime. After feeding the dogs and serving the final walkers their tea and scones, we'd sat down to tea with the children and decided that we would take Joe and Kate to the allotment and see how they fared running together.

'Who's comin'?' I said as I cleared away the plates from the table.

'Me, me, me, me,' said Annas, always keen to be out and about.

'There's a bit of walkin' tha knows,' I told her, and she nodded enthusiastically.

The others were all occupied; Raven doing homework, Reuben, Miles and Sidney crafting something in the workshop and the smaller girls hellbent on having a campfire in the small walled-in garden known as the graveyard behind the old chapel (woodshed) and cooking some of the freshly sprouted rhubarb.

'Right, that's fine,' I said. 'I'll tek Nancy too if you lasses is gonna be playin' wi' fire.'

Clive went to get the quad bike, and I put a beaming, grubby-faced Nancy in the backpack, still nibbling on a biscuit. Even though Nancy was now walking and very mobile she still loved being carried, getting a bird's eye view and experiencing the sights and sounds of life on the farm.

Evenings like this had to be seen to be believed. Tranquillity descended after all the people had passed through and gone. Smoke from the chimney curled and wound its way skywards in a perfect corkscrew, not a breath of wind interrupting its journey. Swallows busied themselves flying at dizzying speeds around the farmyard. We were on the cusp of summer now and everything was alive and vibrant.

'Come on,' I said to Annas, 'let's gan.'

'Tek yer camera,' Clive said from the quad bike, 'it's a bonny neet. I'll gan an' let dogs out.'

I tramped back to the farmhouse for the camera and returned to find Clive standing by the gate out of the sheep pens, and beside him the small figure of Annas, in a peach linen sundress and wellies, both with their backs to me.

'Got it,' I shouted.

Neither of them responded until I got closer, and then Clive turned with a look of abject horror on his face.

'Whassup?' I asked.

'Joe,' he said. ''Ee's dead.'

'*Dead*,' repeated Annas, sounding more excited than upset. Death, through a four-year-old's eyes, was merely a temporary inconvenience.

'Eh . . . dead?' I couldn't quite compute how or why Joe would be dead. 'But you've just fed 'im, 'avent yer?'

'Aye, I did an' all, but I mun't 'ave shut t'door on t'kennel reet an' he's hung 'is daft sel.'

It was a horrible sight. I ushered Annas away, and she ran off at full speed to inform the others of Joe's demise.

We could only guess that, after finding his kennel door ajar, he'd gone for a mooch about and stuck his head through the bars of the gate that led to the sheep pens. Probably the smell of sheep drew his attention. The stoop had two hinges, on which the gate was suspended, and two small lynch pins prevented the gate being lifted off them. It was on one of these tiny metal pins that Joe's collar had become caught. He'd likely pulled and pulled and then panicked and scrabbled his way through the bars to eventually end up at the other side of the gate. Unfortunately, his nylon collar had remained snagged and so twisted even tighter, exerting an intolerable amount of pressure on his windpipe.

It was appalling to see his lifeless body suspended from the gate, his swollen tongue lolling from the side of his mouth, his paws scraped raw and bleeding from scratching at the concrete in desperation. It was wholly upsetting to think that all of this had been going on just yards away from the farmhouse, whilst we were eating our tea. His body was still warm, it had taken just thirty minutes for Joe to die.

'What we gonna do?' I said, not thinking clearly, as there was nothing that we could do.

'Gis' a hand to loosen him,' said Clive.

I am ashamed to say that I couldn't; the sight filled me with revulsion. Of course, I see death amongst the farm animals and accept that where you have livestock you will inevitably end up with deadstock. But this was just something else. I cannot, hand on heart, say that Joe had ventured into the realms of a real companion, he had not been with us for long enough to forge that bond, but I still saw it as an acutely personal loss.

'I cannae see,' Clive muttered as he struggled to unbuckle Joe's collar.

Then, all of a sudden, with a dull thud Joe's lifeless body dropped to the ground.

Clive pulled angrily at the collar that was still caught in the hinge. It came off with ridiculous ease and he flung it aside.

'Erm, I's thinkin' tha' we might need that,' I said, picking up the collar from where it lay by the pen wall. I was sure that sheepdogs would be covered under the farm insurance, so it was important to keep evidence of what had happened. I also realized that, rather like a crime scene, I should perhaps have taken a photograph of Joe's predicament.

Clive picked Joe up, carried him back to his kennel and laid him in his straw-filled dog bed. 'Bloody latch,' he said, as he shut the kennel door behind him. 'I just can't believe it.'

The other dogs were quiet. Usually the noise of the quad bike and the sound of footsteps would signal an outing and trigger a chorus of barking, but not on this occasion. Even Bill was silent, tail at half mast as he looked through the weld mesh towards Joe's kennel. Clive let Kate out and talked to Bill as he passed the kennels.

'Well, Bill, it looks like yer comin' outta retirement, my old friend,' he quipped.

Clive and I set off to the moor, one dog down. We didn't say much, just the occasional comment.

'He were doin' all right, weren't he?' I said.

'Nay I'd won with 'im,' said Clive stoically.

We gathered the sheep up and put them to the moor. Leaving the bike at the bottom, we both walked, Clive with his hands behind his back and his head bowed. Nancy was weighing me down, and I stopped to catch my breath before casting Kate out in a wide arc, to pick up the sheep from the far side of the gutter that bisected the steep rough allotment. It was a big ask for Kate; that evening she was doing the job of two dogs, but pleasingly she worked at a distance with ease and gave me no reason to instruct her further. I stood on a rocky outcrop that gave me the best vantage point and watched her working, her black figure weaving quickly and purposefully through the seaves. Even when she was hidden from sight you could roughly tell where she was according to which way the sheep were running.

Seeing Kate round up the sheep reminded me of being a child, sitting at the kitchen table with my grandad and watching as he did the Pools and the 'Spot the Dog' competition in the newspaper. Every week there'd be a black-and-white photograph taken at a sheepdog trial, showing the sheep, and the handler, but no dog. All you needed to do was put a cross in exactly the place where the sheepdog was to win a prize. He never won. I wondered if I'd be any better at it now.

By the time we got back to the farmyard, the mourners had arrived. The children had been alerted to what had happened by Annas and had come to have a look for themselves.

'Poor Joe,' said Violet, while Edith looked sadly at his body; he was lying on his side as though sleeping in his dog bed.

'Can I dig 'im a hole wi' t'digger?' asked Reuben enthusiastically.

I glared at him. 'He's not even cold yet, Reuben,' I said crossly. 'Anyways, I think that we'll need a post mortem for t'insurance.'

'Aye, yer mebbe right,' Clive said. 'I'll talk to 'em in t'morning.'

Yes, the insurers needed a veterinary certificate, but it was a bank holiday weekend and a dead dog didn't exactly qualify as a medical emergency.

'What we gonna do with him?' I said. 'It isn't good keepin' weather, Clive.'

Unusually, for a bank holiday, the weather was sunny and hot. Perfect for picnics, swimming and attracting hordes of visitors for afternoon teas but not so good for a corpse. Already, I had seen Sidney – with a mischievous glint in his eye – entertaining the visitors on several occasions. They often made small talk with the children and would ask what animals we had on the farm and whether they could see them.

'Yis, we've got lots of animals,' Sidney would say. 'Sheep, horses, cows and chickens.'

The innocent enquirer would then be informed that none of the above were actually in residence at the moment . . . but they could come and look at the sheepdogs if they wanted!

Sidney would show them Roy, who would bounce up and down, and Kate, who would come quietly to the front of the run for her nose to be scratched. Bill would have none of this friendly interaction business and would look at them aloofly from the back of the kennel. And then the highlight of this guided walk – in Sidney's eyes at least – was when he would get to Joe's kennel and could point out his mortal remains, shrouded in a downgraded bedsheet and laid out in his dog bed.

When I realized what he was up to, I warned Sidney that people really did not need to see that!

'The insurers 'ave found a vet that is willing to look at Joe for us,' Clive said, 'on Monday afternoon. You've just got to take Joe to t'surgery. I know nowt about t'fella other than he'll be there all afternoon.'

'Well, I don't suppose that it matters whether he's a good vet or not really,' I mused.

I was going to have to take a drive out towards Carlisle, but it would be worth it not only in monetary terms – working sheepdogs don't come cheap – but also because then we could do the right thing and give Joe the burial that he rightfully deserved. Sidney decided that he would come with me, so Clive was left in charge of serving afternoon teas whilst Edith and Violet constructed a home-made grave-marker out of discarded slats from the woodshed. Reuben got busy on the digger.

I followed the directions I'd been given and eventually pulled up outside a modern building on a quiet street. Surprisingly, the car park beside the surgery was packed and it took a few circuits to find a space for the pickup. In the reception area I found a sea of people, all dressed up to the nines, having some kind of party. Feeling conspicuous in my wellies and torn leggings I pushed my way through the gathered throng.

The crowds parted as, like the opening scene from *The Lion King*, I carried aloft and at arm's length (by now Joe wasn't smelling too pleasant) a shrouded object that was clearly the corpse of a dog.

Sidney melted into the crowd, undoubtedly recounting to all who would listen the tale of woe that had led to us being there.

Sam, the veterinarian, put down his glass of Prosecco as I laid Joe carefully on the table in an anteroom and unwrapped him. Outside the room, I could hear raucous laughter and the sound of music, the mood entirely at odds with how I was feeling upon seeing our sheepdog laid out stone cold on the slab.

I went through the whole horrible incident again as Sam carefully pulled back Joe's lips to look at his teeth and gums and then gently moved his head backwards, parting his fur and inspecting his neck.

'I can see that strangulation has occurred,' he said as he went to the sink to wash his hands.

'Is that it?' I asked. He did seem to be stating the obvious and I felt a bit disgruntled that nothing more scientific had gone on.

'Yes, I'll send a certificate through in t'post,' he said. 'An' a bill.'

'I'll see myself out,' I said as I rewrapped Joe in the sheet, but my words fell on deaf ears as Sam had picked up his wine glass, left the room and rejoined the revellers.

It was a sad end for Joe. We duly buried him in the garden alongside some of the great dogs that had gone before him. He never had the chance to shine, didn't get the opportunity to show us what he could do, but he will certainly be sorely missed – though perhaps not by Bill.

9

Wild Things

The children have always loved to set out on an adventure to the moors and, with a bit of planning, it is possible to combine work with pleasure and take them and a dog or two on a jaunt. Taking a packed lunch and flasks, we would aim for nowhere in particular, just somewhere new, a place where we'd imagine that no person had ever set foot. It always came as a disappointment if we found some kind of man-made mark that told us that we were not the pioneers that we envisaged but were merely following in the footsteps of others who had gone before.

It was usually Bill who would accompany us on these missions, as he did have the ability to at least partially switch off from sheep. Pippen and Chalky would also come but in a low-key manner, following a short distance away, never in a straight line, just haphazardly snuffling around in the undergrowth hoping to put up a rabbit or hare.

With the purchase of The Firs, it felt like we had broadened our horizons even further; now there were new places to explore and more stories to be uncovered. Lonin End lead mines were within close proximity, and although they sounded exciting, it could of course be a potentially dangerous place. The footpath from Keldside, where the ruins of the smelting house stood, followed the River Swale upstream, past The Firs and along a narrow jagger cum drovers' road. Walled on both sides, this trackway was just wide enough to accommodate the horse-drawn carts that transported the lead ore on the journey of a mile and a half from the mine to the smelting house. At any one time, there were fifty men working in the Lonin End mines, some of whom would have lived in a row of cottages that formed part of the site. The rest would have lived within walking distance, passing The Firs along this route every day. If they needed to know the time, they could look up at a primitive sundial on the east wall of the house, now so weathered that only in the brightest of sunshine are the roman numerals on its surface visible. Hewn into the stone above the kitchen window that looks onto the footpath is the word 'DAIRY' so maybe this served as a shop window to take advantage of this passing trade.

An open mineshaft remained at Lonin, three hundred feet deep and only partially covered by a rusting pipe. The children found it fabulously interesting and terrifying, the bigger ones commando crawling to the very edge to peer down into the darkness whilst the little ones hovered at a safe distance.

It was an area that stirred emotions, the last skeletal remains of a once-great industry that dominated the area, employing men, women and children who worked both in the mines and smelting mills. I'd read numerous accounts detailing life in Swaledale two hundred years ago but, as we stood on the site, I just could not visualize it. The overriding feeling was one of serenity and peace, which seemed entirely at odds with the human toil that had shaped the view. The spoil heaps, the remains of the engine house, and the walled pens that once would have held ponies, were all still there to see but slowly crumbling as nature reclaimed what was rightfully hers.

In the past, water and the threat of flooding was a constant worry, as the miners went ever deeper below the surface in search of the lead ore galena, pushing themselves to the very limits of what was achievable with the most basic of tools.

At Lonin End, water was not just the problem but conversely it was also the solution. Birkdale Tarn, a small natural stretch of water nearly a mile away, was extended and dammed. Its location, high above the mine workings, meant that by channelling the water through a series of hand-cut races and pipes, the pressure of the water could be used to power an engine and two waterwheels to pump water out from the mine's deepest levels. It was a marvel of innovation; the eighty-horse-power engine was bought from Ashton Green Colliery and then, for the final part of the lengthy journey, dragged by a team of seventeen horses to its final destination.

Birkdale Tarn is always a picture of loveliness, an open expanse of water that appears to almost sit on a windswept plateau like a giant puddle. Its shallow waters gently lap the shore, shingle to the north side and blackened peat haggs around the remainder. It is invisible until one has almost stumbled upon it and is little known, with few visitors. Its isolation gives it a melancholy air that does not lend itself to anything other than quiet meditation. When the wind whips up and waves sweep across its dark expanse it seems that at any given moment its banks could break, sending a torrent of water down the steep embankment that lies just out of sight.

Throughout the winter months, I'd always look at the frozen rivers and ice-encrusted waterfalls and dream of blistering sunshine, evening swims, nights spent camping out under the stars, and al fresco meals eaten around a campfire. After the damp squib that was the summer of 2017, I was determined to make the most of any good weather that came our way in 2018. The high pressure built up gradually and brought weeks of scorching sun and blue skies. This meant parched fields and rivers running dry, a phenomenon that everyone remembers from their childhood, but which has become a rarity in recent years.

Hats, gloves and scarves were consigned to the cupboard and swimming costumes and trunks retrieved. As the hot spell wore on, the swimming attire became skimpier until the little ones were to be found in the water wearing only pumps. It was the beginning of a summer of wild swimming. Every day after tea we would set out in search of new pools in which to paddle and swim – a procession of bathers laden with towels, snorkels, an inflatable unicorn and, on one occasion, a canoe that began taking on water almost as soon as we launched her. Reuben and Miles had 'liberated' her from her normal – albeit unusual – moorings, resting on the crossmembers of a redundant barn that had once belonged to Tot.

'Does ta think that Tot ever canoed?' Reuben had asked

as we trudged through the fields with the orange canoe resting on our shoulders.

I couldn't see Tot partaking in any kind of watersports. It was far more likely that it had washed up in the river and, being a collector of things, he'd stashed it away for a rainy day . . . just in case. With so many tales of lives lost in the past due to drowning whilst crossing swollen rivers, there was no wonder the older generation had an aversion to water.

For Raven, these nights were more than just evenings spent with the family playing in the water – they were therapy. Now seventeen, she was dealing with the pressure of her schoolwork and impending exams. The sensation of being immersed in the water in secluded and peaceful surroundings was food for the soul, and it washed away her daily worries and stresses. In the case of the little ones, these baths were useful for removing the daily accumulation of dirt and grime picked up around the farm. Raven had always lacked confidence in the water, her school-led swimming lessons at Richmond baths having been cut short after she had a particularly vicious flare-up of psoriasis as a child. The chlorine in the water had dried her skin and left her with weals and sores that she scratched until they bled. Wild swimming was the answer; the water contained none of the nasty irritants that caused her problems and her psoriasis cleared up. Perhaps it was the exposure of her skin to the sunshine, maybe the peaty water itself, but whatever it was the regime worked wonders.

I had wondered if she would be reluctant to go wild swimming, after a desperately sad incident the previous summer. One afternoon when the children were itching for the holidays to start, the school bus carrying the bigger ones did not appear at the usual time. Clive and I did not worry as timings did vary and due to it having been a glorious day there would undoubtedly be tourist traffic out and about on the scenic roads of Swaledale, hindering the school bus on its return journey to Ravenseat.

Time went on, I busied myself making tea, and then the phone rang. It was Rachel, my friend from Bridge End, some three miles away. There'd been an accident, she said, nothing to worry about regarding the children but the road was blocked owing to the air ambulance being in attendance.

Finally, an hour later than usual, Raven, Reuben and Miles returned.

'What happened?' I asked.

Raven told me that she'd planned to swim at Wain Wath waterfalls at Keld that evening with her friends from school but that there'd been an accident there.

'Not another broken leg!' I said flatly.

Every year someone would jump from the top of the falls and do themselves a mischief. It really was a leap of faith if you hadn't taken your time to investigate the depth and the river bottom. Boulders move frequently and to dive or jump without checking would be very risky.

'No, I think that the lad's dead,' said Reuben.

'He jumped in from the top,' said Raven, 'but couldn't swim, got trapped under a rock and was under for eight minutes.'

'Oh my God!' I exclaimed. 'That's terrible – tragic.'

'They got him out, and someone went to get the defibrillator from Keld but he was blue, he wasn't breathing,' she said, her eyes glassed over.

I wondered whether seeing someone's life ebb away in front of their very eyes would leave a lasting impression on Raven, Reuben and Miles, because it had obviously shaken them, but I was surprised at their resilience. I questioned Raven a few days after about how she felt about the incident as her longstanding dream was to become a doctor, and I felt sure this might have put her off. I need not have worried, for it actually did entirely the opposite and spurred her on, but it did make all the children mindful of what can happen if one does not take care.

My aim was for the children to be confident swimmers and to be able to recognize the dangers. I'd make a point of checking the depth and current; in some places where Whitsundale Beck narrowed the water could easily have swept the little ones off their feet. So, while the younger children paddled and played on the banks, the bigger ones swam or jumped and dived depending on the depth of the water.

Violet is terrifically sporty and had only recently begun to have swimming lessons at school; not so long ago she had set off to school excitedly with her swimming costume and towel in a bag.

'How did swimming go?' I'd enquired at tea time.

'Great!' she said enthusiastically, then a perplexed frown developed on her face. 'But the water was in an 'ouse.'

I laughed; it had never occurred to me that she would think the swimming lessons would be in a river. Then she went on to describe in great detail the traumatic discovery she had made when in the pool: a plaster.

'Ugh, it was 'orrible,' she said, wrinkling her nose.

I smiled thinking of all of the undesirable 'things' that might be found in and around the outdoor places where we swam. Slippery green river weed, the fronds of which were coated with orange algae, grew on the rocks along the waterside and around the waterfalls. The boys would insist on throwing great globs of it at each other as it had great sticking-power. Corpses did occasionally appear: a sickly sheep would develop an unquenchable thirst prior to taking its last gasp, go to the water's edge to lap the water and then topple in, only to be washed up further downstream. Pippen and Chalky would patrol the riverbanks while we swam; the children believed that it was loyalty that kept them pacing back and forth, but it was more likely to be the presence of the other residents. If there ever was a scuffle or a brief sighting of one of these furry inhabitants, I would describe the creature as looking 'like a water vole' which sounded far more endearing than a rat.

In fact, Raven had been vehemently complaining about rats recently, though her grievance was to do with the rising costs of buying them.

'Fifteen bloody quid for a dead rat.'

'What d'ya need one o' them for?' I'd asked.

'For school. Rat dissection club.'

When we glimpsed a very large rodent, it was more difficult to pass it off as being anything other than a rat, though fortunately encounters with these sizeable terrors were a rarity.

'Maybe it was an otter?' I said to Sidney as he described a sighting of a rat that he was adamant had been brazenly sunning itself on a rock. Sidney narrowed his eyes with a look of disbelief on his face and shook his head.

'Naw, Mam,' he said, 'it weren't no otter.'

''T wasn't,' Edith chipped in, 'cus I've seen yan, a real un.'

Edith and Violet started reminiscing about an otter that resided at Reeth, a village further down Swaledale, some fifteen miles downstream from where we are at Ravenseat.

'That's amazing,' I said, heartened to hear that there had been otter sightings in the locality. 'I've nut ever seen yan.'

It surely must mean that there'd been an upturn in the state of the river if it had enough small trout to be able to support the resurgence of a mammal that had once been rare.

'It were massive,' said Edith, her arms outstretched to their maximum length, an exaggeration for sure, I thought.

'An it 'ad big feet, I'll tell ya,' said Edith enthusiastically. 'Paws, I mean.'

'It'd bin run over,' added Violet sadly.

'Whooah,' I said. 'It'd bin run over?'

Edith and Violet nodded in unison.

'Bobby brought it to school wi' him for show and tell,' said Edith in a very matter-of-fact manner.

This amused me no end, not the death of the poor otter, obviously, but the fact that the school had accepted this very macabre offering for the children's weekly discussion. I couldn't imagine such a thing happening in a suburban comprehensive.

One fine summer's evening, we decided that we would paddle, swim and clamber our way down Red Gulch Gill towards the washed-out ruins of the old drovers' footbridge that had once spanned Whitsundale Beck. This bridge had at one time allowed the safe passage of animals from one side of the valley to the other without the need for a long detour via Ravenseat. Unfortunately, like many other tiny little bridges, it had fallen into disrepair and eventually collapsed, and now only the square-built stone foundations at either side of the beck remained. En route, we encountered the steep-sided cliffs and narrow ravine known as the Boggle Hole, a place that had always intrigued the children with its vertiginous drop into darkness and the echoes of rumbling water that resonated from its depths. It seemed likely that, long ago, the children from Ravenseat, who would have walked past it on their way to Keld school, had named it after the boggle or ghost who surely lived there.

Days that we could gain access to Red Gulch Gill were few and far between as, even during a dry spell, the journey involved wading through the water. It was too challenging for the little ones, who'd unhappily have to sit this adventure out and stay at the farm with Clive. The rest of us did not know what we might encounter on the half-mile or thereabouts trek, that was the beauty of it. It was surprising how a river that had fashioned its route through solid rock could suddenly change, its path altered by a rockfall or landslide. One morning at lambing time, I'd walked through the Black Howe pasture, picking my way along the trods and through the sheep looking for any new arrivals, only to stare in sheer disbelief at the steepest part of the land which, only the previous day, had consisted of nothing more than short cropped turf ridges and the occasional boulder. Now, it was a mixture of clay-streaked soil, exposed roots and bare rock. Overnight, a landslide had taken a whole slab of ground from the hillside and deposited the slump of upturned turves and mud in the beck bottom below. Tons of soil stripped away in the blink of an eye. I couldn't imagine quite how it would have seemed to an onlooker. Further upstream, a tributary, Hoods Bottom Beck, looped and twisted its way towards Whitsundale Beck. After a particularly torrential downpour, the river broke its banks and the water found itself a more natural course, leaving behind a stagnant oxbow lake. It was all a geography lesson unfolding in front of our eyes.

My intrepid entourage were in high spirits as we made our way through the Close Hills pastures. Stopping for a breather by the barn known as Miles' Cow'as, we agreed that the evening was perhaps the most beautiful that we had ever seen. The trees that lined the far side of the river were a glorious medley of green. I took a photograph, but it did the scene no justice; to appreciate it you had to be there, to listen to the silence and watch the almost imperceptible quiver of the leaves in the lightest of summer breezes. A more sublime view you could not have wished for. We scrambled down the Black Howe pasture in our swimming costumes and pumps, vaulted the low wall and crossed the beck. Then we set down our towels on a grassy knoll beside the first sharp bend in the beck before it cut through the gorge. We ambled alongside the water's edge, the bank narrowing until there was no place to gain a foothold. The girls and I waited whilst Reuben and Miles slipped cautiously into the water, stumbling around, unsure of the riverbed.

'Yer fine. The bottom is rough, mind,' said Reuben, 'but not too deep if yer stay at the side.'

A trench ran down the middle and here, due to the deeper

waters and stronger current, the beck ran cold, though still bearable. The girls waded on down into the darker ghyll, the sides of the ravine now only around twenty feet apart, though the near-vertical cliffs rose some hundred feet above us at either side. Saplings clung to rocky outcrops and on every overhang, where rivulets of water streamed down, were small tapering buds of mineral deposits, the beginnings of stalactites.

'Fossil!' shouted Violet. 'I've found a fossil.'

She was now running her hand over an enormous slab of sandstone that lay half in, half out of the water. The impressions of hundreds or maybe thousands of cockle-like shells covered the abrasive surface. We moved on. The route was now more precarious as we had small waterfalls to scale and were climbing solid rock now, worn smooth by the constant flow of water over thousands of years. The stone had been sinuously carved and layers of striated colour were visible under the surface. The mellowing evening sunshine was reflected on the clear surface of the river. The sight was breathtaking in its simplicity.

We sat at the top of the cascade, our legs dangling into the water, and looked up at the strip of blue sky far above. Occasionally, there'd be the sound of a solitary bird flapping its wings as it exited its nest, disturbed by our presence but, no matter how hard we looked, the echoes reverberating from the cliff walls confused our ears and we could never catch a glimpse.

As we sat there, fingers and toes dabbling in the gently flowing waters, I told the children a tale about a local shepherd. The story goes that in 1894 whilst out tending his flock near The Firs, James Iveson, who lived at the hamlet of Angram, had met with a freak accident. He'd sat down on a boulder beside Sledale Beck for a moment's rest, just as we had done, and reached out to touch another huge boulder beside him. And then, in his own words: 'Suddenly, without the slightest warning, it spun round – though it must have been some forty tons in weight – and crashed up against the rock on which I was sitting. It caught my leg below the knee, cutting the muscles, smashed my strong boot to pieces and tore the leather legging.'

James, obviously in excruciating pain, tried to free his leg but without any success. Realizing that he was trapped there and unlikely to be found before he bled to death, he called his faithful sheepdog Bess to him. He searched through his pockets for a pencil and paper, but all he could find was an envelope addressed to himself, which he duly attached to her collar and told her to 'git away hyam.' Bess was reluctant to leave him and, in the end, he had to throw stones at her to drive her away.

The children were loving this real 'Lassie' story. It was four hours before Bess arrived home, which I thought was impressive.

'Imagine if it'd bin Chalky,' I added. 'I wouldn't like to 'ave 'ad to rely on 'er to instigate a rescue.'

A search party of fifteen men set out. Ten of them, after searching for hours, gave up. But the remaining five had Bess with them and finally they found the stricken shepherd at ten o'clock that night, seven hours after his accident. His rescuers then had to go back and get chisels and hammers to release him from the vice-like grip that imprisoned him. He was finally carried up the bank to a sledge and back to the safety of Stone House Farm some twelve hours after his ordeal began. He was on crutches for thirteen weeks and, although he eventually recovered, he was left with a permanent limp. Even though time and tide have moved the original boulders, the place where this incident occurred is still known locally to this day as Iveson's Trap.

That kind of folk tale was what the children, and I, too, for that matter, really loved. It was history that you could relate to and, to some degree, reach out and touch. Stories of the everyday common people, people that left their mark on the landscape by having places, fields or barns named after them. These places remained long after their namesakes had departed the world, permanent memorials albeit with no physical epitaph, just an everlasting story passed down from generation to generation by word of mouth.

We lost track of time that evening. When we finally reached the remains of the stone bridge the moon had risen. For a few hours we had been in a different world, a place entirely devoid of any human mark. Unfortunately, in order to get back to where we abandoned our towels, we had to retrace our steps and, with dusk approaching, the Boggle Hole was now living up to its supernatural name. We part-swam and part-waded back, now cold and shivering as the light had faded. I was very grateful for the children's companionship on our trek to such a divinely beautiful but wholly unsettling place.

The children talked of little else the following day, being in the water in the moonlight and what they'd seen. To the casual listener it must have sounded like a trip into the wildest, remotest place on earth rather than to the bottom of the Close Hills pasture less than a mile away from home.

It might sound like I have an aversion to going places, but the truth is that we were extremely busy. The everyday commitments that came with running the farm, the shepherd's hut bed and breakfast, the afternoon teas, plus the added work that came with The Firs meant that every day was a busy one. Everyone had to pull their weight and help out at whatever level they were able. The older ones were happy to earn pocket money by helping out with visitors and guests, the younger ones just came along and joined in the melee. The fact was that in modern terms we 'worked from home', thus the children would be involved in whatever activity was going on, whether it was baking or washing, or feeding the dogs or hens – there was the mundane to contend with as well as the exciting.

I would lurch from gliding down to the picnic benches with an afternoon tea balanced on a tray to apprehending a sheep that required urgent medical attention. In fact, it was supremely difficult to keep the realities of farm life separate from that of afternoon teas. Very early in the tourist season we'd had an unfortunate incident when one of our elderly yows had been struck down and overcome by a sudden urge to try and squeeze out her innards in full view of the paying guests at the picnic benches. I must admit that both Clive and I had been warned by the eyes and ears of Ravenseat – Sidney – that there was something afoot in the Low Bobby Dale but, upon inspection, we decided it was just a 'show'. A little prolapse of the rear end that would, hopefully, rectify itself if left well alone. Needless to say, it absolutely didn't, and by the time I realized that something needed doing sharpish, I was already besieged with customers. Raven took over service.

'Sid,' I hollered to the little ginger-haired figure dancing about on the garden wall. 'Ga an' get me a harness an' prolapse spoon – now!'

There was no time to waste, this was serious. Things in the gynaecological department had escalated to nearly the point of no return . . . literally. All I could do was crouch beside the stricken yow and, after returning her lady bits from whence they came, hold them in place with one hand, my knee on her neck keeping her still and on the ground.

Sidney had disappeared into the barn to get what was needed and a little group of people had now taken his place, though rather than dancing on the wall they leaned over it.

''Ow do,' boomed a voice across the field. ''Ave yer baked today?'

I smiled, and said, 'Yep', all the while feeling a bit uncomfortable about both mine and the sheep's position. 'Raven'll sort yer out,' I added in the hope that they'd disappear. No such luck. And where was Sidney? It seemed like he'd been gone for ages.

'Ah's wondering if you'd sign mi book,' shouted the figure, holding it aloft.

I pointed out, politely, that I was currently not in a position to come and sign anything.

'D'ya scones 'ave currants in 'em?' he enquired, settling in for a chat.

Returning a prolapse to its rightful place was something that any shepherd worth his salt should be perfectly adept at and, although it was not a task I enjoyed, I was always rather proud of my handiwork but not so much so that I needed an audience.

Fortunately, occasions such as that are still few and far between but, when planning a visit to Ravenseat, it's best to accept that a working hill farm will firstly be inaccessible, secondly be subject to adverse weather conditions, and thirdly is highly likely to be populated by animals.

I pitied the poor visitor I found sitting rigidly at the picnic benches one day.

'There aren't any birds around, are there?' she asked worriedly.

I explained that there really were no guarantees that there wouldn't be. Ravenseat was renowned for its population of wild birds and there was the added issue of Miles's chickens, which were totally free-range, not to mention our bug-eyed peacock who often put in an appearance. We concluded that, for someone with a bird phobia, there would be far better places to enjoy an afternoon tea than at our farm.

There was a perpetual cycle of scavengers patrolling the picnic benches and the customers were never without a pair of pleading hungry eyes peeking up at them begging for handouts. Usually it was the terriers. Pippen, now elderly, had a pronounced limp that appeared to manifest itself exclusively when in the company of diners.

'Oh ma Gawd . . . has your dawg been abused?' drawled one American who had been completely taken in by Pippen's bluff.

Chalky, on the other hand, would just launch herself onto the table, deftly pick up a scone in her jaws and then make off with it, only stopping to eat her ill-gotten gains when safely out of reach. The solution to this irritating behaviour was to shut the terriers in the stables while we were serving afternoon tea but, unfortunately, this left open a window of opportunity for the peacock. With the terriers out of the way, he would strut his way down to the benches and circumnavigate the picnic area with his high-stepping, exaggerated gait. He would home in on the currants in the warm scones, his favourite, although he was also partial to the little pots of whipped cream that accompanied them. He really had no fear and would approach anyone, rapidly blinking, his head cocked to one side in an enquiring manner. Then he'd help himself to whatever was in range rather than what was on offer. The peacock absolutely knew when I was after him. Sometimes, I'd just snarl at him and chase him back up into the farmyard, other times I'd send a small missile in his direction. I'd invariably miss, but for a while he'd abandon his position.

If the peacock disappeared, this made way for the chickens. They gave him an especially wide berth, having become fed up of his amorous intentions towards them, but once he was out of sight they would close in on the rich pickings to be found by the picnic tables. Last, but by no means least, were the most successful and accomplished scavengers of all: Clemmie and Annas, recently joined by Nancy. Many a happy hour would be spent relieving walkers of their sweets and crisps. It was nothing short of a miracle if there was no form of hijack taking place.

Without realizing it, our scavengers have contributed to a zero-waste policy at Ravenseat. Food-wise, nothing goes to waste; there are hungry mouths to feed whichever way you look. Tea is drunk out of china cups and requests for take-out cups are rebuffed as the paper and plastic objects would just cause litter issues. If you have not got time to stop, then keep walking. Even so, one environmentally conscious Texan walker did manage to find fault.

'Hey, liddle lady,' he said. I was already irritated, and he'd barely started.

'I see that you have an open fire,' he went on, walking uninvited right into the farmhouse. 'Are you not worried about emissions and the atmosphere?'

I was tempted to ask haughtily about the purpose of his trip but the fact that he was clutching a Coast to Coast guidebook answered that question. It turned out he was on a two-week walking holiday, which sounded like it ticked all the boxes regarding green credentials but was less impressive when you factored in the luggage being picked up and dropped off by a van every day at each stopover and the flights to and from the US of A. I have more reasons than many to want to leave this world in a better state, in order that my children can inherit a good, healthy planet. We try to raise a family that is mindful of the countryside, nature and the environment. We do not take foreign holidays, in fact we don't take any holidays. We recycle, plant trees, build ponds and expend a great amount of our time trying to enhance the landscape in which we live. But what is achievable or even possible in rural areas differs vastly compared with urban areas. It is all just a matter of doing your bit.

Epilogue

A Day to Remember

Quite soon after The Firs was ready, in September 2017, we had some farmers come to stay. Dairy farmers from West Cumbria, they were looking for a change of scenery but not too far away as they still needed to be on hand should one of their herd of Jersey cows calve. They had everything arranged, a relief dairyman would see that the cows were milked morning and night, feed the calves and also keep them updated via the telephone. As a keeper of livestock myself, I know that it is never a simple task to leave the animals in somebody else's hands, however capable they might be. There is always a niggling worry that something might be overlooked. But still, my guests were happy to be taking a few days out from their busy farming schedule and, after I'd given them the guided tour of the house and showed them the facilities, they presented me with a gift, a large bottle of Jersey milk.

I looked approvingly at the plastic bottle filled with what I could already see was the creamiest of milk. It brought back memories of my younger days, working on dairy farms and in milking parlours. A perk of the job was that I was allowed to take home a couple of pints of milk every night. I'd decant it from the cooling tank into a stainless-steel can after the evening milking. Unpasteurized and unsterilized, it tasted indescribably wonderful – ambrosial even – wholesome rather than insipid.

'It's fresh out of the tank today,' said one of the two brothers whose families were staying. 'Put it in t'fridge and it'll keep.'

I doubted that it would, actually – he didn't know about the vultures that hovered around the fridge at home – but I told the children that it would be in their best interests to leave the Jersey milk alone. A rice pudding fit for a king would be theirs if they could manage not to succumb to temptation.

As it happened, the entire family became distracted by a major trauma that was playing out in the Keldside Allotment, a piece of land we have between The Firs and Ravenseat. Two autumn calving cows were in there, grazing amongst the molinia, wind grass and heather. We hoped that they would calve before the weather turned, that the calves would be born outside on the clean ground, have the chance to find their feet and get a little autumnal sunshine on their backs before they were all brought into the barn for the winter. Every day we would go and have a look at them, taking a couple of canches of hay with us to keep them happy, biddable and sweet.

'Coosh, coosh,' we'd shout, and the two roan cows would limber towards us.

One drizzly morning, we had only one cow. The chances

were that the other had made away to find a quiet sheltered place to give birth. Clive and I both walked up the wallside, where delicate gossamer webs draped between the seaves now glistened with droplets of dampness. We plodded on, upwards through the pasture where the wall dipped and rose, weaving a sinuous trail through the heather and haggs. It didn't take long before we found a gap in the wall. The bottom row and foundations were still intact, but the rest of the stones had rolled and tumbled to the bottom of the scree that formed part of the adjacent ghyll. We studied the scene carefully.

'Yer don't think that t'cow's got summat to do wi' this?' I said, wiping my nose and staring at the stones and surrounding ground looking for hoofprints.

'Nah,' replied Clive. 'She'll have calved and be laid up quiet somewhere with her calf.'

We agreed that, whilst we were there at that moment, we should repair the wall. I went over the collapsed section and down the screes to retrieve the stones whilst Clive made a start on putting the wall up. It took little time, and was not the neatest job in the world, but it was stockproof. Carefully, I clambered back over the wall, stopping briefly to sit astride it. From that vantage point I could see both of our cows, one at the bottom of the field quietly eating hay and the other previously missing cow striding purposefully out through the heather in the opposite direction from where we were standing. Beside her, a small roan calf half-trotted and half-stumbled, its tail aloft. The pair made away into the distance until they disappeared over a hill end and out of sight.

'She'll calm down in a day or two,' Clive said. 'Her blood's up, best thing for her is to leave her well alone.'

I agreed. There was nothing to be gained by hunting her down, the calf was clearly mobile and strong enough to keep up the pace. We would get to see whether we had a heifer or a bull calf in a day or two, we just needed to be patient.

The following day there was, once again, only one cow awaiting the arrival of hay expectantly. I walked up the field to the same vantage point from where I had seen the other cow the previous day. It was a clear, dry day and you could see for miles. Startled grouse chattered as they took off and then glided to a safer distance; flocks of young lapwings danced and swooped, shining silver under the bright sunshine. They'd soon leave these bleak hills when the cold winds began to blow, the first sign of the changing seasons and the impending winter. I climbed a little higher, sure that I would catch a glimpse of my cow but admitted defeat at the brow of the hill that overlooked the second deep gully. Tomorrow would be just fine. It wasn't the first time that a cow had gone into hiding after giving birth, as it was a natural thing to want to do – all animals did it to some degree. In a small field, or in a barn, there are limitations on privacy but out here she had room, space and freedom, and that was all good, I reasoned.

Day three was almost the last day of the summer holidays and we all went to look for our missing cow and calf. Nancy was in the backpack, and Clemmie was being towed along by Edith, the mother hen. The other cow showed no concern whatsoever for her absent friend and carried on eating hay, dolefully watching us work out our search-and-rescue strategy. We made sure that everybody had an area to cover. The quad bike would have made the task much quicker and easier but unfortunately it could not traverse the steep slopes of knotted, wiry heather.

It was like looking for a needle in a haystack, with so many undulations, ghylls, cascading becks and deep rills where mother and calf could be hidden from view. For every natural obstacle surmounted, another appeared ahead; it was never-ending. The children played and explored, slid down screes, rolled down grassy slopes and then got hungry and fed up.

Then came the shout.

'I've found the cow!' bawled Reuben, dancing a little jig on a craggy stone-topped slope. He pointed furiously downwards and then tipped his head on one side whilst gesticulating a sort of sawing motion across his neck. The children now all set off running, one by one disappearing from sight. By the time I reached the scene of the incident, there was a crowd of onlookers all surrounding the body of my missing cow. I was panting, Nancy now weighing heavy in the backpack.

The cow lay on her side, at the bottom of a steep ravine, her torso bloated, skin taut, legs akimbo. Her neck was outstretched, her head bent backwards, her pallid tongue lolled to one side, her eyes open but lifeless. Water flowed around her. It was a shocking sight.

Sidney moved in close, edging sideways through the water. He gently put his hand on her side as if to detect any warmth. There was none to be found; she had been dead for a couple of days.

It was quite inexplicable how she should have lost her life in such peculiar circumstances. It appeared that she had fallen down into the ravine, though this alone would not have been enough to kill her as she seemed to have little in the way of physical injuries. Staggers or milk fever might have caused her to become unsteady, or maybe she suffered heart failure. Whatever had happened, the result had been terminal.

Miles and Edith, having recovered from the shock of this discovery, were now examining the cow's corpse at close quarters. Violet had gone one step further and was sitting on the cow, perched on its side, looking downwards, deep in thought, her mittened fist resting on her forehead. The macabre picture resembled Rodin's *The Thinker* sculpture.

After a moment or two, she said brightly, 'Mam, where's the calf?'

I confessed that I didn't know.

'Yer don't think that she fell *on* it?' said Miles.

'That woulda been kinda unlucky,' said Raven phlegmatically.

We had a good look and concluded that there was no evidence to suggest that there was anything under the cow other than silt, mud and stones, and that the calf must be somewhere else. The search resumed, only this time for a small, newborn calf weakened through hunger and almost certainly frightened. We knew that time was against us and that we wouldn't have long before he or she succumbed to either starvation or the elements.

We started our search in the vicinity of the calf's dead mother, radiating slowly outwards. We combed the banks, hunting high and low. The chattering of the children would surely rouse any creature that was lying low to avoid detection.

A couple of hours passed and other than finding in the peat the pickled and preserved ancient tree root that had once grown there, and a half-deflated foil helium balloon that had become tangled up in the heather, we had nothing to show for our efforts. The children were flagging, their spirits at a low ebb, and finally we gave up and set off back home.

'Ne'er mind,' I said cheerfully. 'He'll be hungrier tomorrow, we'll stand more chance of finding him.'

I hated walking away with only half of the mystery solved, though was remaining positive, maybe overly so, but I was not ready to concede defeat just yet.

The next day dawned bright and we decided to set about the search more tactically. The smaller family members would stay at home with Clive rather than hinder us in our search, the rest of us would take our lunches so we could stay out longer and hopefully we would be victorious.

A whole day was spent traversing the ghylls looking for any sign of our lost calf, such as hoofprints and droppings (although the fact that there could be nothing going in at one end meant that there'd be little produced at the other). Once again, we returned down-spirited and empty-handed. In my mind, it was looking like the worst-case scenario: that the calf was dead, and we would never find him.

Days went by and every day we looked. The children went back to school midweek but every evening we'd have the same discussion at the tea table.

'Ave yer found t'calf?' Reuben would enquire in between mouthfuls of dinner.

'Nope,' I'd say.

'Mi mate has a drone, he can bring it at t'weekend,' he said.

It was certainly worth a try and, on the Saturday, we resorted to aerial shots, viewing the Keldside remotely from above via a drone. There was nothing to see.

'That's it,' I announced on the Saturday evening. The last and only confirmed sighting of the calf was almost a week ago; its mother had been dead for a minimum of five days. We had exhausted every possibility and we just needed to draw a line under the whole sorry incident. 'The calf is dead, there's no point looking any more.'

We got on with our jobs as usual. Although it was a special day, being my birthday, the overriding feeling was one of failure. Edith, as bright as a button, announced at lunchtime that she was taking the binoculars and was going to have one final look across into the Keldside to see whether she could spot anything. I nodded and smiled sadly as she skipped off dressed in her boiler suit and woolly hat. She came back to report her failure an hour later.

Edith's enthusiasm was in direct contrast to my despondency and, by the time tea was on the table, I had decided that a pessimistic mindset was no good at all and was badgering Clive to come with me for one last look around the pasture. He muttered and chuntered and reluctantly agreed to come along. We dressed warmly, as the light was already fading, and then set off, sitting side by side on the quad bike, to the field gate. A biting coolness in the evening air numbed my face and I pulled my hat down over my ears and my scarf over my nose. Clive prattled on about cows, the trouble they caused and how on earth we were going to retrieve the dead cow's body. I wasn't listening.

'Let's ga reet up onto t'top amongst ling,' he said, 'neet'll be 'ere soon.'

We walked together for a little while, then separated. After a half-hour of haphazard wandering, we met up and made for home.

'Reet, that's it now, no more lookin',' said Clive firmly.

The cragged hills across the valley were momentarily backlit by a fiery crimson sunset. I shivered and pushed my hands deeper down into my pockets. Clive came over and put his arm through mine, and we walked together saying nothing. In many respects we were quite different in our outlooks. I would analyse and look back to see how I should have done things differently, Clive would look ahead and never bother to cry over spilt milk. The combination of these two approaches was probably the best way to be.

The dewy dampness of evening now wet our leggings as we walked through the heather, and every so often one of us would stumble as our wellies were caught up in the wiry heather roots. We were nearly back to the field gate and the quad bike when suddenly, without any warning whatsoever, a small roan calf sprung up immediately in front of us. Quite who was most startled, me or it, I could not guess. For a fleeting moment, the calf stood, unsure of what to do next, and Clive, always a quick thinker, leapt upon it before it had the chance to make a bid for freedom. The calf let out a weak and tired bawl as Clive, now on his knees, wrapped an arm around its neck and stopped it from escaping. The calf clapped to the ground, its legs folded underneath it as it sat. It stared fixedly ahead with fearful eyes that had sunk back into the sockets.

'Pick it up,' panted Clive, still holding the calf down. 'It's as light as a feather.'

I bent and gently picked it up, its long limbs dangling as I put my arms underneath it.

'My God, I just can't believe it,' I said in a mixture of jubilation and disbelief.

The dazed calf rested its head on my shoulder and mooed, the faintest and most pathetically tired noise I had ever heard. I swear that even Clive was lost for words.

'The children are gonna be beside themselves when I show them,' I said.

We clambered back onto the quad bike, and I held the calf tightly across my knees, its legs either side of the seat. We were now full of excitement and enthusiasm for our next challenge, to very carefully nurse our calf back to health. It was a miracle that it had survived so long without sustenance and now we needed to very carefully reintroduce it to milk, without it being too much of a shock to the system.

'We gently need to warm . . . her,' I said, taking a quick look under her tail. 'It's a heifer. We need to bring her inside an' get 'er up to t'fire.'

We disembarked at the front door and carefully brought the calf into the porch. The children, now dressed for bed and lounging around in the living room, were astounded when confronted with the calf who now stood forlornly in the middle of the room with her head hanging downwards and ears drooping.

Edith covered her mouth and burst into tears.

'I know, I can hardly believe it myself,' I said, 'but now the hard work begins, we must feed her, lal' and often.'

Now, as the little heifer stood surrounded by adoring children dancing a jig on the stone-flagged floor, it became clear just how malnourished she was. Her guts were concave, hollow on either side below her hip bones which jutted out awkwardly and angularly. We were going to have to be very careful with her nursing.

'What milk are you gonna give her?' asked Clive. 'Needs to be good stuff, she cannae stand a tummy upset.'

I wasn't so sure that I knew the answer. Milk-replacement powder was not going to be good enough, but maybe full-fat blue-top out of the fridge would suffice. Colostrum wasn't required as she had been suckling on her mother until the accident; there was absolutely no way that she could have survived for this duration if she had not had the critical first feeds of this energizing, antibody-filled milk. I went to the fridge, opened the door and looked to see what I had that might be suitable. And there it was, a large plastic bottle full of Jersey milk, the most nourishing wholesome milk a weak calf could ever have.

'What's her name gonna be?' I asked the children as she hungrily sucked the milk from the bottle. I reminded them that, as she was a pure-bred Shorthorn heifer, we would be keeping her to join our small herd of cows and that she needed a pedigree name that began with J. The cows were named alphabetically according to their year of birth; that way we could always tell how old they were by their name.

For once, everybody agreed. She was to be called Joy.

Joy was reared on the bottle as a pet, and her uncertain start in life proved to be of no detriment as she thrived and grew into a strong and beautiful cow. She did not live alone; her companion was a shorthorn bull calf that went by the name of Jimmy. He too was reared on the bottle, though not because he was an orphan but because he had a thoroughly detestable mother. Such is life, there is no accounting for the wills and ways of animals, and Jimmy's mother showed no maternal instinct whatsoever, refusing point blank to even acknowledge her newborn. She would spin around when he attempted to suckle, cavort about and then flatten him against the wall at any given moment. Eventually, we admitted defeat and reared him ourselves. It suited Joy to have a companion, as it did at least give her some notion that she was, when all is said and done, still a cow.

The children were devoted to their care for six months until it was time to wean them, and Jimmy and Joy were just as devoted to the children. It was a terrifying sight to see the children setting off on a bicycle ride, a raggle-taggle peloton accompanied by two half-grown calves galloping and bucking alongside, letting out an occasional cow-kick here and there.

A whole year later, and Joy and Jimmy had been reintroduced to the other young cattle and were living in the high pasture at The Firs. It really struck me how much of an impression the animals had made on the children's lives when the bigger ones returned from feeding them their daily ration of cake.

'OMG,' said Raven, in typical teenage speak, when they all returned. 'Like, I can't believe it.'

'What?' I said. I was ever so hopeful that the penny had dropped – perhaps they did know what day it was after all.

'I took a picture of Joy and Jimmy and all the other cows at Firs looking over the wall,' she carried on.

'Mmmmm,' I replied.

'It's a special day, Mum!' she chirped.

Now this was looking promising, maybe she had remembered.

'It said on mi phone that on this day last year we found Joy! It's her birthday!'

I have to say that I was a little crestfallen, though not surprised, that Joy the cow's birthday had ranked as being more memorable than mine, as we have never really been much into celebrating the passing of time. For the children, time is not seen as such a precious thing, merely wished away with hopes and dreams of what the future might hold. The summer holidays last forever and for the less-scholarly children the school years drag on endlessly. Nothing changes and that continuity is really what bestows upon us a sense of belonging and purpose.

We do all seem to live in a smaller, more-connected world today, but for us here in the highest reaches of the Dales, where the bronzed heather-clad hills dominate the skyline and the silver-streaked rivers flash beneath expansive leaden skies as they have done for centuries, the close-knit community and the demands of the changing seasons are still at the heart of what we do. To be able to have the world at our fingertips if we so choose is a marvellous thing indeed but, as more and more visitors pass through our farm, then maybe we should realize just how lucky we are to call this little corner of Yorkshire our home.